CHILDHOOD DEPLOYED

Childhood Deployed

Remaking Child Soldiers in Sierra Leone

Susan Shepler

NEW YORK UNIVERSITY PRESS

New York and London

NEW YORK UNIVERSITY PRESS
New York and London
www.nyupress.org

References to Internet websites (URLs) were accurate at the time of writing.
Neither the author nor New York University Press is responsible for URLs that
may have expired or changed since the manuscript was prepared.

Library of Congress Cataloging-in-Publication Data
Shepler, Susan.
Childhood deployed : remaking child soldiers in Sierra Leone / Susan Shepler.
pages cm
Includes bibliographical references and index.
ISBN 978-0-8147-2496-5 (hardback) — ISBN 978-0-8147-7025-2 (paper)
1. Child soldiers—Sierra Leone. 2. Child soldiers—Sierra Leone—Reintegration.
3. Children and war—Sierra Leone. 4. Sierra Leone—History—Civil War, 1991–2002.
I. Title.
UB418.C45S44 2014
362.7'7—dc23   2014002554

New York University Press books are printed on acid-free paper,
and their binding materials are chosen for strength and durability.
We strive to use environmentally responsible suppliers and materials
to the greatest extent possible in publishing our books.

Manufactured in the United States of America

10 9 8 7 6 5 4 3

Also available as an ebook

CONTENTS

Figures

Tables

AFRC	Armed Forced Revolutionary Council
APC	All People's Congress
BECE	Basic Education Certificate Examinations
CAFF	Children Associated with the Fighting Forces
CDF	Civil Defense Forces
CPA	Child Protection Agency
CRC	Convention on the Rights of the Child
DDR	Disarmament, Demobilization, and Reintegration
ECOMOG	ECOWAS Monitoring Group
ECOWAS	Economic Community of West African States
FAWE	Forum of African Women Educationalists
FHM	Family Homes Movement
FTR	Family Tracing and Reunification
ICC	Interim Care Center
ICRC	International Committee of the Red Cross
IDP	Internally Displaced Person
INGO	International Nongovernmental Organization
IRC	International Rescue Committee
MSWGCA	Ministry of Social Welfare, Gender, and Children's Affairs
NCDDR	National Commission for Disarmament, Demobilization, and Reintegration
NCRRR	National Commission for Rehabilitation, Resettlement, and Reconstruction
NGO	Non-Governmental Organization
NPRC	National Provisional Ruling Council
RUF	Revolutionary United Front
RUFP	Revolutionary United Front Party

SLA	Sierra Leone Army
SLPP	Sierra Leone People's Party
SLTU	Sierra Leone Teachers' Union
UNAMSIL	United Nations Mission in Sierra Leone
UNHCR	United Nations High Commissioner for Refugees
UNICEF	United Nations Children's Fund

I taught mathematics in a rural secondary school in northern Sierra Leone as a Peace Corps Volunteer from 1987 to 1989, several years before the brutal civil war began. When I returned to Sierra Leone ten years later in October of 1999, my first task was to try to understand the war. It seemed impossible to me that the country I had known (and loved) could be home to such horror. I had of course followed the war from the United States, and tried to discover what happened to my friends and "my village." But news was scarce, and biased, focused on the capital city, Freetown, and mainly on counting up bodies and atrocities. How could I make sense of it? What methods could I use to try to get a handle on this traumatic event, particularly on the postwar experiences of the children who had been forced to become combatants? I knew that it required more than summing up the facts, and it required moving out of the capital city. It meant talking to people of all kinds of backgrounds to achieve some sort of tapestry of meaning out of their sometimes conflicting but always compelling stories.

I was often told over my two years of fieldwork in Sierra Leone, "this war is deep." In other words, the war is complex, not completely understood by observers or even participants, with many secrets yet to be revealed. In this book I can only give a taste of the complexity, and I will by necessity focus on aspects of the war that are important to understand for my argument regarding the experiences surrounding child soldiering.

Anthropologist Valentine Daniel says in the introduction to *Charred Lullabies*—a highly personal account of the conflict in Sri Lanka—that writing his book arose in part out of a need to tell the stories that people had entrusted to him.

Stories, stories, stories! I have never known for sure if I am their pris-
oner or their jailer. . . . In opening up my tape recorder and notebook to
my informants, I took upon myself the responsibility of telling a wider
world the stories that they told me, some at grave risk to their lives, only
because they believed that there was a wider world that cared about the
difference between good and evil (Daniel 1996, 5).

I listened to hundreds of stories about the war. Everyone had a
story; everyone had felt the war's impact. One of the first tasks was to
understand a kind of shorthand that people used to refer to different
events of the war, for example January 6th and May 25th. (Everyone
knew what these dates meant. It was not necessary to add the year.)
But despite a certain shared language, I realized that there is no such
thing as *the* war. It, whatever *it* was, moved through time and space
in such a way that its participants came to understand it differently
based on where they were living, their tribe, their gender, their class,
and often their blind luck. There is no way I or anyone could tell the
story of *the* war.

War is a total social phenomenon, affecting not just the combatants,
but every person, thing, social structure, and ideal. It plays itself out
not just in the realm of extraordinary physical violence but also in the
realm of symbols, in language, in witchcraft, in the everyday. War is
experienced and narrated in different ways, and all of the layers are crucial in any attempt to understand social trauma at such a scale (Ibrahim
and Shepler 2011).

This book offers, therefore, a partial reflection of the civil war in
Sierra Leone. I experienced it first-, second-, third-, and tenth-hand. All
of these "hands" involved stories about the war, but those stories also
become ways of talking about the nation, gender, tribe, and other categories of social life. War is continually lived and re-created for purposes
that extend beyond the end of the conflict. When I was there, in the
course of my research and daily life, I too told my stories, and in doing
so was part of the postwar world. But rather than just recounting what
happened, this book investigates how people lived through and talked
about "the war" after the fact, the social practice of meaning making,
and how, in particular, they made sense of child soldiers.

ACKNOWLEDGMENTS

With so many debts incurred, I hardly know where to begin. Sincere appreciation is due to so many.

To my colleagues at Gbendembu Secondary School, who were my friends, fellow palm wine drinkards, and first informants before I knew what the term "informant" meant.

To graduate school friends at Berkeley who helped me out of my ill-fitting identity as a mathematician and into my new identity as a union organizer and social theorist: Eric Hsu, Lily Khadjavi, Concha Gomez, and Jim Freeman. And of course, to my "tribe": Steven Hillion, Jordan Tircuit, John Fox, Mary Crabb, and Mark Dolan.

To friends in Sierra Leone, old and new, who welcomed me home after a decade away, despite their own difficult circumstances: M. T. Bangura, Sonny Joe, Michael Kamara, Shellac Sonny Davies, Frances Fortune, Pinkie McCann-Willis, Nell White, and my students and colleagues at Milton Margai College of Education and Technology. To all the child protection workers who tolerated my incessant questions at Family Homes Movement, Caritas Makeni, the International Rescue Committee, Children Affected by War, and elsewhere. To the Sierra Leonean communities who made me welcome, and in true Sierra Leonean fashion asked me to share in the little they had as they were rebuilding from chaos. To the many young men and women who revealed something of their lives to me.

To Jean Lave, much more than a dissertation advisor, a model for how to live a committed intellectual life. That dinner at Chez Panisse, the night before I blithely left for a war zone, meant more than I can say. To other professors who profoundly shaped my thinking and the contents of this book, Gillian Hart and Mariane Ferme.

To a community of scholars who engaged with me and my ideas in various rough drafts around Berkeley. In the Graduate School of Education, Mary Crabb, Raymond June, Amanda Lashaw, and Pam Stello. In the Reconstructing Communities in Crisis Working Group, Robin Delugan, William Hayes, Joseph Nevins, Krisjon Olson, and Susana Kaiser. In a lifesaving dissertation-writing group, Aaron Bobrow-Strain, Nitasha Sharma, and Rebecca Dolinhow. My fieldwork and write-up were funded by a number of generous organizations. The Rocca Memorial Fellowship for doctoral research in African Studies, from the Center for African Studies at the University of California, Berkeley, allowed me to travel to Sierra Leone for two rounds of fieldwork, even when the war was officially ongoing. The American Association of University Women, the Institute on Global Conflict and Cooperation, and the Harry Frank Guggenheim foundation all provided valuable dissertation fellowships.

Other scholars of Sierra Leone, of child soldiering, and of youth and conflict have been mentors or coconspirators; included among them are Aisha Fofana Ibrahim, Susan McKay, Kristen Cheney, Ibrahim Abdullah, Bidemi Carrol, and Doug Henry.

In addition, American University and the School of International Service have provided a wonderfully collaborative home and support, financial and otherwise, for my ongoing research in Sierra Leone. Thank you to all my colleagues in International Peace and Conflict Resolution, and to a number of research assistants including Janelle Nodhurft, Sharyn Routh, and Cate Broussard. To my "Sohos": Rachel Robinson, Brenda Werth, Kristin Diwan, Kate Haulman, Adrea Lawrence, and Elizabeth Worden.

Finally to my husband, Wusu Kargbo. I couldn't have done it without you.

Introduction

Jerihun was the site of an internally displaced persons (IDP) camp between Bo and Kenema in the Eastern Province of Sierra Leone. When I arrived there in 2001, the camp was fairly new, having only been in operation for about six months. It was designed as a transit camp for Sierra Leoneans returning from refugee camps in Guinea, mainly Kono people who had been away from their villages for ten years. The camp housed several thousand IDPs in small stick and mud huts built by the occupants themselves. They were completely supported by international nongovernmental organizations (NGOs). All of their food, water, education, medicine, and other supplies came from the United Nations High Commissioner for Refugees (UNHCR, the United Nations refugee agency), and other subcontracting NGOs. I chose to visit this camp because former child soldiers were being reunified with distant family in the camps—sometimes, ironically, with family they had never met. My goal was to study the reintegration of these former child combatants and their experiences with both Western agencies and their home communities after the war was over.

After clearing my presence with the NGO running the camp, I walked around and greeted people. The NGO staff there was unprepared to help me and, in fact, seemed completely unaware of who the ex-combatants were or where they were staying. I heard from various residents that things were hard at this camp, that in Guinea they had been much better supplied. More to the point, the Mende people in the neighboring village wanted them out (as a ploy, they surmised, to get more money from the white men for hosting refugees in their locale). There were also complaints that the Sierra Leoneans working at the camp were stealing supplies and that the camp occupants did not have any access to land or tools. This meant that the occupants mainly sat around the camp and waited. They were waiting for their home areas to be declared officially safe and then for UNHCR to take them back. As a result of these problems, the camp was emptying out. I thought that might mean that people were going back to their homes, but I was told, "No, they are looking for other camps." One Mende woman from Bunumbu asserted that she would like to go back, but there was nothing there. She told me that she had passed overland from Guinea and had actually traveled through her hometown of Bunumbu on her way to the camp. When I asked why she had not stayed then, she said, "There was no food, no buildings, what would I do? So instead I sit in the camp all day."

After asking around for children who had taken part in the war, I met three "child ex-combatants" at Jerihun Camp. They all had similar stories: abducted approximately seven years prior at around age seven. When I met them they were fourteen, though to me they looked younger. They told me that they had spent their years with the rebels as porters and general help, not really as combatants. They said they never *fired* guns, though they would often carry them around and follow after the big men, explaining "even rebels like big man business." That is, the rebel commanders gained prestige for having a number of young followers around them. One of the children, Sahr, indicated that he had received "training," meaning some form of military training.[1]

Sahr is a common name among the Kono people, meaning firstborn male child. Sahr took me to meet his "brother"—actually a member of his extended family whom he had never met until coming to the camp. Since he could remember his village name and the names of his parents, the child protection agency caring for him after demobilization was able to make the connection. Sahr's brother said he and his family had been in Guinea for about eight years until they were brought back to Sierra Leone by UNHCR. They had to take in the boy, he explained, because although they did not *know* him, he is family to them. (Also, there is now an extra name on their feeding card at the camp.) Sahr's brother had put a lot of effort into making their little house very nice, even planting a flower garden with seeds he brought with him from Guinea.

Sahr had never been to school. At the age of fourteen he was in class 1, the equivalent of the first grade in the United States. I asked him if he minded being in class with little kids. He said, "No. Everything has its stage." I asked him if he could write his name and he proudly replied, "Yes!" I asked him to write it for me in my notebook. He went to his brother for help with the "S" but his brother refused, urging, "No, you can do it yourself." So he wrote for me proudly SAHR—all in capital letters, with a backward S.

*　*　*

This book is about the reintegration of former child soldiers in Sierra Leone. The international community defines a child soldier, or a "child associated with an armed force or armed group," as "any person below 18 years of age who is or who has been recruited or used by an armed force or armed group in any capacity, including but not limited to children, boys, and girls used as fighters, cooks, porters, messengers, spies or for sexual purposes. It does not only refer to a child who is taking or has taken a direct part in hostilities" (UNICEF 2007, 7). In Sierra Leone, children were recruited and used by every fighting faction; girls as well as boys were trained to fight and to carry out a full range of other war-related activities. Since 2002 when the decade-long civil war in Sierra Leone came to an end, some forty international and local nongovernmental organizations have worked there to reintegrate an estimated seven thousand former child combatants (DeBurca 2000; Coalition to Stop the Use of Child Soldiers 2002).[2]

This book is about what happened after these child soldiers demobilized and struggled to return to "normal" life, rather than on their mobilization into fighting forces or what they did while they carried guns. How did they conceive of their time "in the bush"? How did other Sierra Leoneans see them? What was the process of so-called reintegration like? This book examines, from the ground up, children's and adults' own experiences of postwar rebuilding. The analysis is based on eighteen months of ethnographic fieldwork in Sierra Leone in interim care centers for separated children, in schools struggling to integrate children whose education had been disrupted by war, in nonformal apprenticeship programs, and in selected communities where former child soldiers have been reunified with their families.

This book is also about childhood: the childhoods of the child soldiers, but also about the modern conception of childhood, forged in the West and exported around the globe via child rights discourse and practice. What happens when Western models of childhood bump up against various local models of childhood in the struggle to reintegrate former child soldiers? In the West, we understand child soldiers using our own models of childhood, framed by institutions such as the United Nations Children's Fund (UNICEF), international nongovernmental organizations such as Save the Children, or transnational advocacy organizations such as the Coalition to Stop the Use of Child Soldiers. These organizations' key proposition is that since child soldiers are, after all, *children*, they cannot be held responsible for their actions. They should more rightly be seen as victims and every effort should be made to protect them and reintegrate them into normal childhoods. This viewpoint is aligned with our commonsense understanding of childhood, namely that children are innocents in need of protection and should be spending their time in schools, not acting as participants in war.[3] But does this view align with the views of Sierra Leoneans? Of the former child soldiers themselves?

By exploring the meaning of childhood, as it is lived, negotiated, and deployed strategically by multiple actors, this book argues several points. In Sierra Leone, "youth" is best understood as a political category. Indeed, "child soldier" as a category is co-created by Sierra Leoneans and Westerners in social practice, not in the experiences of individual children. The reintegration of former child soldiers is, in many ways, a *political* process having to do with changing notions of childhood as one of the central structures of society. Struggles over childhood and child rights in postwar Sierra Leone are productive sites in that they become the locus for all kinds of other political struggles. Like feminist scholarship, which can generate insights into the broader structures of society through a focus on the micro-politics of gender, this work of childhood studies reveals the broader structures of society through a focus on the micro-politics of age. Close attention to how reintegration differs for boys and girls, ex-combatants of different fighting factions, and formal and informal reintegrators, illuminates the contours of these political struggles. In this book, we hear the voices of the former child soldiers themselves, in their multiple social contexts. The most innovative contribution of this work is that it addresses the vast majority of former child soldiers who forego participation in formal reintegration programs and, in the language of NGOs, "spontaneously reintegrate" after war.

Moreover, this book argues that UN- and NGO-sponsored programs for child soldiers have unintended effects as they seek to change the very nature of youth as a political category in Sierra Leone. NGO activities purporting to help former child soldiers are in some ways buttressing the old-fashioned patron–client relationships at the heart of the corrupt postcolonial state and disabling prewar forms of youth power. Sierra Leonean former child soldiers find themselves forced to strategically perform (or refuse to perform) as the "child soldier" Western human rights initiatives expect in order to most effectively gain access to the resources available for their reintegration into normal life. These strategies don't always work; sometimes Western human rights initiatives may ultimately do more harm than good.

This book provides answers to the obvious question, can former child soldiers return to normal life after unspeakable violence? What does their reintegration look like? What works and what does not work for former child soldiers, both in their own terms and in the terms of the communities into which they are reintegrating? The practical conclusions are that programs for former child soldiers work best when they work within local models of child protection, for example through child fosterage and apprenticeship, rather than through excessive institutionalization and reliance on Western child rights–based models. Ultimately, this book concludes that an ethnographic approach to understanding children's actual lived experience can contribute to more effective policy and programming that will help to support the "best interests of the child."

Childhood Studies and Situated Practice

The Western model of childhood is most clearly articulated in the 1989 United Nations Convention on the Rights of the Child (CRC).[4] In the

West, this model is almost common sense:[5] children are innocent, children should not work, children should be in school, children should live with their families, and children should be allowed to express themselves. Within this framework, child soldiering is wrong because it contravenes the notion of an ideal childhood. This version of childhood did not appear out of the blue; it has a history and politics of its own. The model of youth specific to the Western industrialized nations was worked out in the colonies and recirculated in the metropole (Stoler 1995) and is now upheld and perpetuated by global institutions such as the United Nations. As senior anthropologist of childhood Jo Boyden puts it, "The norms and values upon which the ideal of a safe, happy and protected childhood are built are culturally and historically bound to the social preoccupations and priorities of the capitalist countries of Europe and the United States" (Boyden 1997, 192). The spread of this global model—that is, its application in parts of the world far from its creation—has to be understood as part of the history of colonialism and of colonialism's offspring, development. Indeed, the spread of the Western ideal could be seen as the colonization of childhood, one of the central ideas of what is now known as childhood studies (James and Prout 1997; Burman 1994; Pupavac 2001; Stephens 1995). Children's participation in war is not a new phenomenon (Marten 2002; Rosen 2005); what is new is the international child protection framework that has constructed the identity "child soldier" where it previously did not exist, through techniques from the fields of education, psychology, and social welfare. How do young people learn to enact or embody the identity "child soldier"? During the war they learn to fight and to survive, and they learn a factional identity, but while "in the bush," they generally do not know the term "child soldier." They learn to apply the term to themselves by going through and between a series of institutions *after* their participation in war.[6]

Often questions about child soldiering have been framed in terms of structure and agency (Coulter 2009, Denov 2010). Practice, or, more completely, social practice theory, is a potential way out of this dead-end duality.[7] What do I mean by practice? First, it is not just the opposite of theory, as in the oft-decried divide between theory and practice; and second, it is not just a description of what people do, as in their actions instead of their thoughts or feelings. Instead, practice theory is a theoretical tradition that allows for a new relationship between mind and body, and between structure and agency. Social practice theory does not place the social in mental qualities, nor in discourse, nor in interaction (Reckwitz 2002, 249). Practice theory can be defined through two distinct but complementary motives or research programs. The first is an empirical program, ethnographic in its sensibility, for understanding social and organizational life. The second is a theoretical one aimed at "transcending perennial problems in philosophy and social sciences, such as Cartesian dualism and the agency-structure problem" (Miettinen, Samra-Fredericks, and Yanow 2009, 1312). That is, instead of conceiving of social reality as made up of structures that individual agents move within and against, practice theory sees social life as the sum of practices, where practices are habits of thought, or action, or body.

Thus, the kinds of knowledge with which practice theory is concerned are located in lived action (competence of acting, style, practical tact, habituations, and routine practices), in the body (gestures, demeanor, corporeal sense of things), in the world (in being "at home" with what one does, dwelling in it), and in relations (encounters with others, relations of trust, recognition, intimacy) (Forester 1999, 102). Practice theory requires describing a "field" in all its complexity, while simultaneously noting that the field itself is made by the practices of social reproduction. "Child soldier" is produced in practice—partially determined by institutional structures, and partially as a result of children's own strategizing—in various social, historically and geographically situated sites.[8]

In order to carry out a practice theory analysis, one's point of departure cannot be the reflexive (and unreflective) condemnation of child soldiering as an egregious child rights violation. Adopting a critical approach allows an understanding of child soldiering from the perspective of social practice. We must ask how former child soldiers and other Sierra Leoneans themselves understand and employ child rights discourse and the construction "child soldier" to serve their own motives. What are the strategies—in Bourdieu's (1977) sense of the word—of children and adults in response to global ideology? Anthropologist of childhood Sharon Stephens argues that more research is needed on how global discourses such as "the rights of the child" are worked out locally, in practice. In her groundbreaking work on children and the politics of culture, she attests,

The crucial task for researchers now . . . is to develop more powerful understandings of the role of the child in structures of modernity. The historical processes by which these once localized western constructions have been exported around the world and the global political, economic, and cultural transformations that are currently rendering childhood so dangerous, contested, and pivotal in the formation of new sorts of social persons, groups, and institutions (1995, 14).

The goal of this book, then, is to uncover the political content of what is often presumed to be the apolitical construction of childhood. The volume compares the experiences of formal and informal reintegrators, boy soldiers and girl soldiers, and children affiliated with two different fighting factions: the rebels (the Revolutionary United Front, or RUF) and the local militias (the Civil Defense Forces, or CDF). Its central question is: How does "modern childhood" function as an ideology? How does modern childhood frame the possibilities for debate and analysis of a range of issues, including youth culture at the local and global level, the war in Sierra Leone, issues of gender and of the postcolonial?

How to Study War: Placing War in Social and Historical Context

The first violence is the decision where to start telling the story. Do we start with the incursion of the rebels into Sierra Leone? With the corrupt system that caused them to rebel in the first place? With the colonial legacy that allowed the corrupt system to come in to force? For some in Freetown, Sierra Leone's capital, the war did not start until it came to their front doors, many years after the first attacks. For others it started much earlier.

There are several ways to break down the various schools of thought about the war in Sierra Leone. There are those who emphasize external causes—the international diamond trade, international gun runners, the rapacious nature of extractive global capitalism, the history of colonial domination and its impact on political forms—and those who emphasize internal causes—such as the breakdown in the patrimonial system, the collapse of the educational system, the corruption of the local elites, and underlying ethnic tensions.[9] Most scholars acknowledge both external and internal factors as important and admit that they are inextricably linked.[10] Thus, the corruption of local elites is due, in part, to the legacy of the colonial system, and illegal international diamond trading is only possible because of the internal breakdown of the state.

The first book-length exploration of the war by an anthropologist was Paul Richards's (1996) *Fighting for the Rainforest*. Richards sees the causes of the war in a general "crisis of youth" pointing to the breakdown of a patrimonial system and the reactions of "excluded intellectuals." He posits rationality, organization, discipline, and calculated visions of social change by a movement that is led by "quite highly educated dissident" intellectuals. Richards says in his conclusion: "I am more than ever convinced that the (Revolutionary United Front rebels) must be understood against a background of region-wide dilemmas concerning social exclusion of the young. . . . (T)he increasing resort to violence stems from past corrupt patrimonial manipulation of educational and employment opportunities" (Richards 1996, 174).

A set of Sierra Leonean scholars—historians and political scientists mainly—have taken issue with Richards's theses (Abdullah 2004a).[11] These scholars support the centrality of youth to any explanation of the war but deny many of Richards's more extreme assertions about "excluded intellectuals." The Sierra Leonean scholars are much more likely to point to internal causes for the war.[12] We can leave aside the various arguments about whether youths were responsible for the war, or whether they were dupes of powerful political forces outside their control. What is important here is to acknowledge the belief, among both outsiders and Sierra Leoneans, that the situation of youth was a central driving force behind the conflict. This analysis fits well with the current popularity of demographic and economic explanations for war, citing in particular the "youth bulge" in sub-Saharan Africa as a cause for much conflict as vast cohorts living in poor economies fail to find work and resort to killing each other. Indeed, in this theoretical atmosphere, policy makers increasingly see youth as a dangerous segment of the population, requiring urgent programming in education and livelihoods (UNDP 2006; Urdal 2004). On the other hand, youth are "the future," and policy makers greatly desire their participation in civil society. A rash of recent book titles lay out the dichotomy: Are youth in Africa *Vanguards or Vandals?* (Abbink and van Kessel 2005), "Makers" or "Breakers"? (see Honwana and De Boeck 2005), *Troublemakers or Peacemakers?* (McEvoy-Levy 2006). Professor of African Studies Mamadou Diouf puts it this way, "Today, young people are emerging as one of the central concerns of African Studies. Located at the heart of both analytical apparatuses and political action, they also have become a preoccupation of politicians, social workers, and communities in Africa" (2003, 2). This means, finally, that understanding the war requires understanding the various meanings of "youth" in Sierra Leone, including youth culture, patrimonialism, and everyday practice surrounding youth.

A Brief History of the War, and the Fighting Factions

The goal of this overview is not to offer a full accounting of the war. Political economist David Keen (2005) and investigative journalist Lansana Gberie (2005) each give an excellent overview, and the report of the Truth and Reconciliation Commission is exhaustive in its detail (Sierra Leone Truth and Reconciliation Commission 2004). Here I merely introduce the main fighting factions and provide a broad sense of the timeline and issues involved.

The Revolutionary United Front (RUF) Rebels

Starting in 1991, the Revolutionary United Front (RUF) made incursions into the south and east of the country from neighboring Liberia, which had been in the throes of its own rebel war for the previous several years. There has been a great deal written about the character and origin of the RUF (Richards 1996; Abdullah 1997; Ellis 1999; Zack-Williams 1999; Peters 2011). The RUF initially enjoyed some support from the population, as there was great dissatisfaction with the prevailing system, and talk of the need for a violent overthrow had been around for a long time in student circles and elsewhere. In my experience, most Sierra Leoneans agreed that the RUF may have begun with a core group of politically oriented revolutionaries, but their activities soon devolved into terror and banditry. They were an unpredictable group with a shared amorphous revolutionary language, but without a well thought out plan of how to achieve their ends. Different commanders had different styles, with the worst overseeing murder, rape, child abductions, amputations, and torture. The RUF sometimes made gestures of solidarity with the people only to turn against them the following day.

The RUF abducted people into their ranks, both children and adults. But in some cases there was a kind of natural association between rebel occupiers and the young rebellious sectors of society. Some Sierra Leonean scholars describe three types of youth involved in the movement: "the urban marginals (or 'rarray man dem'); . . . the 'san-san boys' (or illicit miners), who live very precarious lives in the diamond mining areas, and who joined the rebel movement in large numbers when mining towns and villages were overwhelmed by the RUF; and socially disconnected village youth . . . who are contemptuous of rural authority and institutions, and who, therefore, saw the war as an opportunity to settle local scores" (Abdullah et al. 1997, 172).

The RUF's leader was Foday Sankoh, a former army member and northerner, previously convicted of participating in a coup attempt. RUF members were scruffy, and lived and trained in the bush. They drew on a superficial pan-Africanism without the associated historical consciousness (Abdullah 2004b, 2002). Their costume included elements of military dress, sunglasses and bandannas, and also, at times, wigs and other elements of cross-dressing (see Moran 1994, for a description of Liberian cross-dressing fighters). They sometimes looted nice clothes and shoes in order to dress well. They smoked marijuana and listened to Bob Marley and Tupac Shakur. Their imagery drew on reggae music, sometimes explicitly critiquing the "Babylon system."[13] The "system" can mean anything from the system of capitalism to the power of local chiefs; it can mean the corrupt educational system or the system of patronage.

The RUF and some portions of the Sierra Leone Army (SLA) joined forces and were known as "sobels": soldiers by day and rebels by night (Richards 1996, 6; Ferme 2001b, 223). This is perhaps not completely surprising since the same types of youth who were drawn into the RUF were heavily involved in the Sierra Leone military as well. According to Sierra Leonean historian Ibrahim Abdullah, the ranks of the government army multiplied more than fivefold during the course of the war.

War came to be regarded as a survival strategy for youth who had suffered high levels of social exclusion (Abdullah et al. 1997, 172).

The Armed Forces Revolutionary Council (AFRC)

In May 1997 the Armed Forces Revolutionary Council (AFRC), an alliance of the RUF and a portion of the SLA led by Major General Johnny Paul Koroma, staged a coup, and the elected government went into exile in neighboring Guinea. The advent of the AFRC made public what many had known or suspected for years: that an alliance existed between the RUF and at least some portion of the SLA. The alliance was based on political expediency and on the continuation of what had become a very profitable war economy based primarily on the export of diamonds (see Reno 1997a and 1998).

The Civil Defense Forces (CDF)

Through the mid-1990s, to people in Freetown, and in the north, the war seemed like a southern and eastern problem. To people in the south and east, it seemed as if no help was forthcoming from the capital. Many of them were living in refugee camps in Guinea and in internally displaced persons (IDP) camps in Sierra Leone. Partly in response to the informal alliance between the RUF and the SLA, the Civil Defense Forces (CDF) were organized (Muana 1997, Leach 2000, Ferme 2001a, Ferme and Hoffman 2004, Hoffman 2011) out of existing hunting secret societies.[14]

Hunting societies are formed around the knowledge required to hunt, namely knowledge of weapons and the ability to move quietly through the forest in search of prey. Historically, hunters have controlled herbal medicines said to render one invisible or even bulletproof. The Kamajohs[15] are the best-known group of the hunting society–based militias that made up the Civil Defense Forces. These were not strictly hunters, but a new kind of fighting force grown out of the hunting society, making use of preexisting secret society iconography and use of local plants for mystical purposes, but essentially a new creation.[16] Children and young men were initiated into the society as fighters in order to help the force grow. The Civil Defense Forces (including the Kamajohs and other ethnically organized fighting groups), like secret societies, have a public face and a private (secret) face. In media images, they often appear dressed in "traditional" locally woven clothes covered with leather charms, but paradoxically carrying very modern arms. They draw on long-standing understandings of the secret societies' control of powerful magical forces in their public representation. However, there is also a more mundane aspect to the forces. When you see a CDF fighter in everyday life, he is usually dressed as any other man—with only a small charm around his neck, sometimes called a "safe," to distinguish him. But even that is not very distinguishing as non-CDF men, women, and children sometimes wear similar charms.

The United Nations Mission in Sierra Leone (UNAMSIL)

In mid-1999, the Government of Sierra Leone in exile and the leadership of the RUF met in Lomé, the capital of nearby Togo, to work out a peace deal. The agreement involved an amnesty for members of the RUF and a power-sharing agreement under which Foday Sankoh would become vice president and minister of mines and the elected politicians would return from exile. After the Lomé accords, the UN sent a military force of its own, the United Nations Mission in Sierra Leone (UNAMSIL). Slowly, UN troops were stationed throughout the country (for example, Kenyans in Kenema, Indians in Mile 91). In 2000, the accords fell apart as Foday Sankoh's followers again became violent, taking UN peacekeepers hostage and violently engaging with protesters in Freetown. Foday Sankoh himself went into hiding, and was eventually arrested and placed into UN custody. In January 2002, President Kabbah officially declared peace, and his Sierra Leone People's Party (SLPP) government was reelected in May 2002. A Truth and Reconciliation Commission came and went in 2003, with most Sierra Leoneans taking little notice. The Special Court has concluded its trials of all defendants, including Charles Taylor, the onetime president of neighboring Liberia. Johnny Paul Koroma is missing and presumed dead, and the two most high profile defendants, Foday Sankoh of the RUF and Sam Hinga Norman of the CDF, both died in custody while awaiting trial.

Today, although most agree that peace has been achieved, there is widespread concern that most of the issues that led to the conflict still remain. Although people have seen what war can do and never want it to return, they are worried that unless something is done to address issues such as corruption, development, education, and the plight of youth, another war could happen in the future.

My Approach

It is necessary to understand the war within multiple frameworks, partly because it is a complicated story and partly because Sierra Leoneans use that complexity as a cultural resource. As education expert Antoinette Errante (2000) reminds us, a postwar period has its own political logic, quite different from that before or during. Child rights are part of the landscape of competing postwar narratives, all partially constitutive of social reality. The way people talk about and frame the war is crucial to understanding how the global child rights discourse is vernacularized in Sierra Leone (Merry 2006).

Of course, research is not a neutral exercise, and, especially in the context of armed conflict, it has considerable potential to infringe on the privacy, well-being, and security of its subjects. Jo Boyden cautions scholars conducting research with children confronting adversity to be aware of informed consent issues, expectations, accountability, the protection of children from harm, and the need for respect for the research subject. She concludes that "ethnographic research, based on a mix of observations (participant, unstructured and structured), personal testimony and other forms of narrative, has an important role to play, not least because of its potential in harnessing children's own understandings and views" (2001, 5).

Despite these challenges, ethnography is still the best approach to studying the lived experiences of children. Ethnography can reveal the life worlds of children from their own perspectives and illuminate alternate indicators of well-being.[17] In my fieldwork I experienced directly the challenges of ethnography with children affected by war. I found that I was most successful in gaining insight into their lives when I did not approach their experiences of violence directly; indeed, questioning a former child soldier about his or her time "in the bush" often yielded one of a set of stock answers, a script I came to recognize well. I found that the most revealing times were those spent simply "hanging out" with children participating in their activities.

During eighteen months of research in Sierra Leone in 1999, 2000, and 2001, I followed social and spatial networks as children moved through the supposed stages of reintegration. Anthropologist Carolyn Nordstrom describes this method as "ethnography of a warzone"—"an experimental methodology based on studying a process . . . rather than a study based in a circumscribed locale" (Nordstrom 1997, 10).

I spent at least six months in various interim care centers for former child soldiers, meeting the staff and participating alongside children in their daily activities. I also investigated formal and informal apprenticeships, foster care arrangements, schools attended by former child soldiers, the meetings of the national Child Protection Committee, and various local and international child protection NGOs. During the second half of my fieldwork I lived in five different communities in which former child soldiers were reintegrating. In addition, many of my most interesting discussions happened in bars, markets, or on public transportation. The presence of a white woman speaking Krio was always a cause for amused curiosity, and after I explained the purpose of my presence, Sierra Leoneans almost always had something they wanted me to hear. Finally, I kept an eye on the local media—radio and newspaper mainly—for representations of child soldiers in the public sphere.

I selected the five primary field sites with a view toward representing various axes of differentiation in the population of former child soldiers.[18] I documented the views and experiences of both the ex-child combatants and the members of the communities in which they were being reintegrated. I designed the research so that I could study a small number of individuals in their social context, essentially a cross section of ex-child combatants in several select communities, rather than performing a statistical study of these groups throughout the society. This represents a kind of purposive sampling, chosen to reflect a range of experiences across a number of axes. In each location, starting from the few former child soldiers identified for me by an NGO worker, I located all of the former child soldiers in each location, many of whom I would not have been able to find through NGO assistance alone. In particular, this method allowed me to identify large numbers of informal reintegrators, a population of child soldiers at that point inaccessible to researchers at UNICEF and other program-based organizations.

Table I.1. Comparison of Field Sites

Name	Region	Ethnic group	Type of settlement	Ex-child soldier population	NGO involvement
Pujehun	South	Mende	Town (district headquarters)	RUF to CDF (Kamajoh)	Small (CAW)
Masakane	North	Temne	Village	CDF (Gbethi)	Small (Caritas) resettlement
Rogbom	West	Temne	Village	RUF/AFRC (short-term)	Small (Caritas) resettlement
Jerihun	East	Mende, Kono	IDP camp	RUF (long-term)	Large (IRC, UNHCR)
Freetown	West	Mixed	Capital city	Mixed	Large

Table I.1 summarizes some of the differences in the five sites. This mix of locations allowed me to compare urban and rural experiences, regional differences, ethnic differences,[19] and factional differences. It allowed me to study both boys and girls, as well as children who were affiliated with the fighting factions for just a short time, and those who spent their entire childhoods in the bush.

Chapter Outline

We cannot understand child soldiers in Sierra Leone without understanding the Sierra Leonean model of childhood and youth. Chapter 1 describes aspects of the practice of childhood in Sierra Leone—child labor, secret society initiations, child fosterage, and education and apprenticeship—that are continuous with the participation of children in armed forces. Sierra Leoneans have their own culturally specific reactions to child soldiering that are not reflected in global child rights discourse. What is disturbing to them is not a lost innocence but a separation from family and training and the idea that the nation loses a generation.

Chapter 2 describes Western interventions in Sierra Leone on behalf of child soldiers: demobilization, interim care, psychosocial activities, schooling and skills training, family tracing and reunification, follow-up visits, and community support. The identity "child soldier" in Sierra Leone is made as young people move through these institutions designed for their rehabilitation and reintegration. Their postwar identities are partly structured by these institutions and partly made in overlapping arenas of social practice as individuals react to and negotiate with the system for their own needs. In particular, children sometimes use discourses of abdicated responsibility—"It was not my fault that I fought with the rebels. I was only a child!"—to help ease their postwar reintegration. But this same notion of child innocence in some ways makes reintegration more difficult, since Sierra Leoneans want children to return to their normal place at the bottom of the social hierarchy.

The "child soldier" is made in and around institutions in multiple and sometimes contradictory ways. The ideological underpinnings of these institutions is a Western, individualistic framework, yet the actual effects are found in Sierra Leoneans interacting with (making and remaking) the institutions in social practice. In Chapter 3 I show how the system-as-designed broke down as a result of the maneuverings of individuals who participated in it in unanticipated and unintended ways that both helped and hindered their "reintegration." Generally, rather than one predetermined circuit from normal life, to the bush, through an ICC and back to normal life, it was possible for individual children to move from any state to any other. Throughout this process the identity "child soldier" became useful in a number of ways.

Many child ex-combatants bypassed the institutions designed for them and simply went home on their own. The child protection NGOs called this "spontaneous" or "informal" reintegration, a sort of residual category for all the children affected by war not participating in NGO activities. Chapter 4 examines the different trajectories of so-called formal and informal reintegrators to further understand how communities organize "reintegration" in the absence of NGO programs. Although the so-called formal reintegrators have better access to various benefits, they must be "out" to their communities as former combatants. This means they cannot use the strategy of secrecy to ease their reintegration and that they sometimes become the target of community anger. So-called informal reintegrators lose out on some benefits, but in general they more easily blend back into their communities than formal reintegrators do. Some informal reintegrators strive to get registered as child soldiers after the fact of their reunification in order to access benefits. Communities collude in this activity, sometimes even fabricating lists of child ex-combatants in order to maximize the community benefits that come with a population of child soldiers. These activities are to some extent undoing the distinction at the local level between formal and informal reintegrators.

Struggles over childhood and child rights are productive sites in that they become the locus of all kinds of other political struggles. Chapter 5 takes up two important distinctions in the population of child soldiers in Sierra Leone: the RUF and the CDF, and boys and girls. The child ex-combatants of the Revolutionary United Front (RUF) rebels and Civil Defense Force (CDF) militias have very different postwar experiences and are understood quite differently by child protection workers and policy makers, even though on the face of it they experienced similar traumas. NGOs argue that because CDF fighters stayed close to home during the war they do not need the same reintegration help as RUF fighters. The other effect of this is that boys of the CDF have a harder time escaping the wartime bonds of membership in locally based militias. This means that children of different factions have different access to the resources of "child soldier," both discursive and material. The CDF boys are in some ways in a worse position than the RUF boys, because the CDF is a hierarchical institution that places young men at the bottom of social hierarchies. Ex-combatant benefits tend to come through still-existing wartime command structures, and in order to access these benefits boys must stay within a patrimonial system rather than adopting the "child soldier" identity based on modern constructions of youth.

Across another axis of differentiation, although many girls were abducted by the RUF—by some estimates as many girls as boys were abducted—they are even less likely than CDF boys to access the benefits that come with the identity "child soldier." Their postwar experiences are quite different from those of boys. Only a handful of girls went through formal demobilization and reintegration programs. There are many reasons for this. Girls are subject to an explicitly *moral* discourse about their participation in the conflict and hence are less able to take advantage of the same discourses of abdicated responsibility as boys. Girls' strategies for reintegration are more likely to include seeking anonymity. The chapter concludes by placing girls' experiences of war alongside the everyday structural violence in girls' lives.

In Sierra Leone, new definitions of youth are being forged in contradictory and extremely political ways. By adopting the modern notion of youth, young people gain one type of political power and lose another. The techniques behind the creation of "child soldier" as a postwar identity have serious and surprising political effects. Hence, the concluding chapter takes up the politics of childhood and youth and extends them to a politics of knowledge creation. It ends with a plea to policy makers and child protection programmers everywhere to take ethnography seriously, and reflect on what a from-the-ground-up critical study can reveal about the impact of their interventions.

1

Youth in Sierra Leone

The romantic model of the child is a profoundly modern
and western construction, emerging in the nineteenth century when industrialization was reallocating the distribution of geographical and psychological space to map onto
the divisions between urban and rural, public and private,
and "home" and empire. These domains were all reflected
in the split between "nature" and "civilisation" that were also
inscribed within the theory of childhood (Burman 1994, 239).

When we in the West think about child soldiers, we tend to do so with
our Western notions of what childhood is and should be.[1] We think we
know what "child" means (under eighteen years of age, innocent, moving through developmental stages, at school, and not at work) and we
think we know what "soldier" means (an adult, well trained and disci-
plined, fighting for a cause or a state), but these words mean something
different from our expectations in the context of the civil war in Sierra
Leone. For a number of reasons, Sierra Leoneans understand child soldiering quite differently than Westerners do.

There are continuities that render the "child soldier" intelligible in the
Sierra Leone vernacular, continuities of practice and discourse. This book
does not claim that the phenomenon of child soldiering is completely
explained by the cultural practices surrounding youth in Sierra Leone,
nor is this an attempt to make reasonable the participation of children
in war. Instead it is an effort to understand historical continuities and
cultural practices and meanings surrounding children and youth that
make the participation of children in conflict somehow legible, and affect
how the category "child soldier" may be used strategically. This chapter
describes how various practices surrounding youth contributed to the
recruitment of children and youth into the fighting forces, and determined, in part, the nature of their participation in the war.

Though the modern "child soldier" is a relatively new phenomenon—perhaps only a few decades old—by now, there are many, mainly
NGO-sponsored, studies of child soldiers.[2] The child soldier studies
almost always begin from a human rights framework (in particular,
a child rights framework), and focus mainly on estimating the numbers involved, recounting individual horror stories, describing the legal
instruments against the use of child soldiers, and evaluating reintegration programming.

To try to end child soldiering, the UN's child protection agency
(UNICEF) and international child protection NGOs such as the Coalition to Stop the Use of Child Soldiers, Save the Children Fund, Christian Children's Fund, Human Rights Watch, and others focus in some
of their literature on finding reasons for it. This literature finds answers
in several types of argument.

It has been suggested that large youth cohorts, so-called youth
bulges, make countries more unstable in general, and thus more
susceptible to armed conflict (Urdal 2004). There is also a simple
argument of supply: in countries where 50 percent or more of the
population is under eighteen, there is a ready supply of children for
recruitment. Various scholars have noted the changing nature of
modern warfare, in which wars are fought less and less by regular
armies and in which civilians are more and more the targets of violence. Children are caught in the middle of both these trends. In addition, as put by the Coalition to Stop the Use of Child Soldiers, "[t]he
widespread availability of modern lightweight weapons enables children to become efficient killers in combat" (Coalition to Stop the Use
of Child Soldiers 2004)

There is also a sense that many children involved in warfare have no
other options, as this selection of excerpts from several child soldier–
focused NGO websites makes clear:

> While some children are recruited forcibly, others are driven into
> armed forces by poverty, alienation and discrimination. Many chil-
> dren join armed groups because of their own experience of abuse at the
> hands of state authorities (Coalition to Stop the Use of Child Soldiers
> 2004).

Others join armed groups out of desperation. As society breaks down during conflict, leaving children no access to school, driving them from their homes, or separating them from family members, many children perceive armed groups as their best chance for survival. Others seek escape from poverty or join military forces to avenge family members who have been killed (Human Rights Watch 2004).

The overwhelming majority of child soldiers come from the following groups: children separated from their families or with disrupted family backgrounds (e.g. orphans, unaccompanied children, children from single parent families, or from families headed by children.); economically and socially deprived children (the poor, both rural and urban, and those without access to education, vocational training, or a reasonable standard of living); other marginalized groups (e.g. street children, certain minorities, refugee and the internally displaced); children from the conflict zones themselves (Coalition to Stop the Use of Child Soldiers 2004).

In addition, NGO literature makes arguments about children and war in terms of the characteristics of the child, in particular that children are easily intimidated and easily indoctrinated:

Physically vulnerable and easily intimidated, children typically make obedient soldiers (Human Rights Watch 2004).

Both governments and armed groups use children because they are easier to condition into fearless killing and unthinking obedience; sometimes, children are supplied drugs and alcohol (Coalition to Stop the Use of Child Soldiers 2004).

There is an element of truth to all of these explanations, what has come to be the conventional wisdom on child soldiers. However, the conventional wisdom is not without problems or ideological biases. The demographic argument begs the question, why then has war not taken place in locations with similar youth bulges? The small arms argument is countered by the reality that in Sierra Leone, for example, most of the violence was carried out not with guns but with everyday tools like machetes and fire (Rosen 2005). With respect to the poverty argument, it is clear that in Sierra Leone it was not only street kids who joined fighting forces. Indeed sometimes family ties led a child to war. Most importantly, the idea that children living in desperate situations will turn to violence ignores the fact that the majority of children did not join in the fighting.

By looking for ways to explain child soldiering, child protection NGOs end up conflating many completely different contexts. Their discourse is about the large scale (war sweeping across the world as a force on its own) and the small scale (the story of an individual child, usually as an illustrative example), rarely making connections between the two scales. By doing this, the NGOs construct the individual powerless child as an illustrative example to bring about pity. These discourses have a universalizing tendency and employ a characterization of children as innocent victims, unable to make their own decisions, objects rather than subjects, needing to be saved by outsiders.

Understandings of "the" child as seen in Western media and NGO representations do not necessarily match understandings of the nature of childhood in a local context (that is, children in some cultural contexts may not be viewed as innocent or weak). Especially in "developing countries," children may sometimes be breadwinners, and they certainly act as agents, strategizing about their lives and situations.[3]

The rest of this chapter unfolds as follows: First I discuss the categories "child" and "youth," in practice, in social relations, and varying across gender, class, and ethnicity. Second, I delve further into four aspects of youth widely studied in West Africa—child labor, child fosterage, apprenticeship, and secret society initiation—and show how in Sierra Leone these practices were continuous with the recruitment and participation of children in the fighting forces. These continuities are based on my own analysis and are apparent to me as an outsider, yet to most Sierra Leoneans they are too taken for granted to be cited as explanations. So, third, I explore how Sierra Leoneans themselves explain the phenomenon of child soldiering through arguments about economic, political, and social breakdown. I look at the history of the participation of groups of young men in political violence of various sorts from precolonial times to the present, a history that gives Sierra Leoneans a ready vernacular for talking about groups of "troublesome" young men. I also touch on Sierra Leonean notions of the nature of youth as inherently controllable, yet liminal and dangerous. Finally, this chapter points to what Sierra Leoneans found horrifying about child soldiers. Their specific distress can only be understood through their own notions of youth, and it has to do with the inversion of age hierarchies, the breakdown of social structures, the effects of improper training and initiation, and the future of the nation.

"Youth" in Sierra Leone: Some Definitional Issues

The main idea of childhood studies is that "youth" and "childhood" vary from culture to culture, and that youth and childhood are categories made in social relations (Ariès 1962; Stephens 1995; James and Prout 1997; Boyden 1997; Malkki and Martin 2003). As Mary Bucholtz puts it, "It is commonplace of much research on youth cultures and identities that the youth category lacks clear definition and in some situations may be based on one's social circumstances rather than chronological age or cultural position" (2002, 526). Even the United Nations definitions contain overlap, with children defined as under eighteen years of age and youth as between fifteen and twenty-four years of age (http://www.un.org/youth). Given these difficulties in defining youth in any general way, anthropologist Deborah Durham (2000) proposes applying the linguistic concept of a *shifter* to the category of youth. A shifter is a word that is tied directly to the context and hence takes much of its meaning from situated use. Likewise, the referential function of youth cannot be determined in advance of its use in a particular cultural context.

The question then becomes: Given the constructed nature of youth generally, how is youth constructed in Sierra Leone? The goal is not to come up with one definition of "youth" in Sierra Leone. Rather, the goal is to show some of the complexities of the category, and to show that it is a category with a great deal of explanatory power for Sierra Leoneans themselves.

Of course, it is impossible to talk of Sierra Leonean perspectives that are untouched by Western notions. Sierra Leone is a deeply globalized place, with a five-hundred-year-long history of Western interaction, including the devastating effects of the slave trade.[4] Concepts of youth have been very much shaped by contact with Western countries, especially England, over several centuries. For example, formal public education in Sierra Leone under the British has operated since the 1780s. Furthermore, the colonial era saw a hardening of the category "traditional," which has important political consequences to this day. Nonetheless, among Sierra Leoneans there is a sense of Sierra Leonean childhood, distinct from Western childhood. So, in the most general terms, a youth is someone who is no longer a child[5] but is not yet a "big man" or "big woman." It is nearly impossible to assign exact age ranges to these terms because of the multiplicity of possible trajectories in Sierra Leone, which vary by gender, region, ethnic group, class, and other considerations. The key point is that the determination of who is a youth is not based simply on age; as elsewhere in West Africa, the transition from child to youth involves certain relations and activities. Some activities are fairly universal; for example, almost everyone goes through initiation into the secret societies that serves as a formal marker between childhood and adulthood. Yet given that, there are a range of different trajectories from childhood and adulthood.

Having said that, there are certain "rules." Generally, youth is over when one marries. Boys usually marry later than girls because they need time to acquire the money and status required to marry. The effect of this is that boys are youths much longer than girls. I told an educated Krio woman I met that I was in Sierra Leone studying changing conceptions of youth. She said, "We don't have anything like that here. A boy is a youth until he is married. A girl is a youth until her first child."[6]

The agricultural economy of the rural areas entails different relations of youth than those in the urban areas.[7] This point was illustrated to me in a presentation by a Sierra Leonean Planned Parenthood worker in Bo. He was giving a presentation to a group of young ex-soldiers about the dangers of early sex. One of the boys raised his hand and said, "Well, in the village, men sometimes get married quite young. Is that wrong?" The man answered something like, "Well yes, it's different than here in town. In the village, once a boy is big enough to cut *banga* [palm kernels] for himself he should marry, but he continues to live with his parents so it's not the same."[8] Clearly, there is confusion even among

Sierra Leoneans, but in general, urban models of childhood and youth are closer to the Western, particularly the British, model than rural ones.

Though it should go without saying that Sierra Leoneans love and care for their children as all human beings do, children in Sierra Leone are also a kind of social security, or as I heard on numerous occasions, *pikin na bank* (a child is a bank). One raises children in part in order to have someone around to take responsibility in old age, so one invests in children when they are young in order to give them the best possible advantages. It is recognized that not all of them will succeed. This is why it makes sense to think of them as investments. It can be risky to sink a lot of capital into one child. The child may die, or end up being "spoiled" by Western education. The child may not be appropriately grateful for the parents' support. Often, parents with many children put their children onto different trajectories so as to broaden the scope of possible outcomes. Some stay on the farm, others are given Western or Islamic education, others apprentice to trades. Girls are married off to other families, so there is less need to invest in them (except of course that they should be made attractive for marriage, and therefore bring the benefits of bride price and linkage with another family). It was often commented to me that one never knows which one of the children will "have a bright star" or will "do better thing" so that it is necessary to invest in all the children in some way or another. One might think of children as something like seeds: one plants them and waters them and does not know which one will grow. In West Africa generally, children are useful, require investment for later payoff, have symbolic value, and are exchangeable.

For this reason, children serve as a kind of prestige marker. Anyone without children in Sierra Leone is thought not to be "serious." One is only fully an adult when one has children. Having children is in many ways more important to the status of adult even than marriage. This is especially true for girls; that is, the time of the birth of their first child is the time when they are generally considered adult. A man with a big fortune but no children would still be seen to be somehow poor.

Certain activities mark one as a youth. Apprentices of all kinds are "youth," no matter their age. One is a youth as long as one is in school, but more boys than girls are in school. Nothing says "youth" like a school uniform. Secondary school students can be in their early twenties, and that figure is rising in postwar Sierra Leone as many whose education was disrupted by the war are going back to start up again after a several year break.

Much of Africa can be described as a "gerontocracy," and it is important that we conceptualize youth as a political category. In West Africa generally, "youth" is constructed in opposition to "elder" and is a political category as much as or more than a biological one (Murphy 1980; O'Brien 1996; Durham 2000). We can thus find "youths" as old as thirty-five years, as, for example, in the Sierra Leone Government's Youth Policy (Government of Sierra Leone 2003). In the introduction to the collection *Navigating Youth, Generating Adulthood: Social Becoming in an African Context*, Christiansen and colleagues put it well:

> As is obvious in current writings on Africa, 'youth' is a highly context-dependent and fluid signifier. But the way we use it in this text, which we think reflects the way it is generally used in West Africa, is as a label for marginalized young (and not so young) people, rather than for a whole population within a certain age bracket. The potential danger of youth is thus not dependent on bulging demographic processes, as popularly supposed, but rather on the number of young people experiencing socio-economic marginalization and powerlessness (2006, 3).

"Youth" in this context is almost equivalent to "subaltern" and is generally understood to be male. Youths are those crowds of people not yet "big men," waiting for their opportunity for advancement, their opportunity to marry and start a family. Ibrahim Abdullah, a scholar of Sierra Leone youth culture, explained to me, somewhat flippantly but still revealingly "in Sierra Leone, if you are unemployed, then you are a youth" (Ibrahim Abdullah, personal communication, April 26, 2005). This definition is reflected in some of the other answers I got to the question "What is a youth?" One young man told me that "a youth can be up to forty, and is anyone who hasn't been given what they should have been given in life, like opportunities or property." After a meeting near my field site, Masakane—at which the imam, the pastor, the women's leader, and the chief spoke—the youth coordinator (a position the local NGO for youth self-development had come up with), frustrated by the lack of youth input, came up to me and said, "You see? They never let us speak!" In a system where elders control the political power, we can understand youth as a political class for themselves.

According to anthropologist William Murphy, scholars often "attribute a benign character to the inequality between elders and youth in traditional African societies. . . . In other words, it is simply a matter of time in the developmental cycle for youth to win independence from elders' authority. This view overlooks the fact that while young men do become old men, not all old men become elders" (1980, 202). He goes on to note that, especially in rural areas, "Consequently, most old men, along with women and young men, remain junior dependents of the elders of the high-ranked lineages. Despite their age they are still essentially 'youth' in their dependence on these elders, and they exercise even less authority than important younger members of the higher ranked lineages" (ibid.). In Sierra Leone today, with life expectancy hovering around forty, one can easily live one's whole life as a "youth."

The so-called crisis of youth is postulated all over Africa, where economic collapse has meant, among other things, that young men are less able to marry so they stay "young" longer (Sommers 2007). Consequently "elders versus youth" is the battle line of political struggle in much of modern-day Africa. The tension between elders and youth is seen by scholars as an increasingly important divide. As the anthropologists Jean Comaroff and John Comaroff put it:

[In Africa, the] sense of physical, social, and moral crisis congeals, perhaps more than anywhere else, in the contemporary predicament of youth, now widely under scrutiny. Generation, in fact, seems an especially fertile site into which class anxieties are displaced. Perhaps that much is overdetermined: it is on the backs of the pubescent that concerns about social reproduction–about the viability of the continuing present–have always been saddled. Nonetheless, generation as a principle of distinction, consciousness, and struggle has long been neglected, or taken for granted, by theorists of political economy (2000, 306).

Of course, those political lines are not new in Africa, but they are taking on new cultural forms in the face of globalization and global youth culture. Understanding the implications with respect to child soldiering, rehabilitation, and the aftermath of war is a major focus of this work.

Aspects of Youth Practices Continuous with Child Soldiering

The model of childhood and youth described above is common throughout West Africa. This section works through four important aspects of youth in the Sierra Leone case: child labor, child fosterage, apprenticeship, and secret society initiation. Each aspect is explored as a set of practices and relations that, in the sense described above, *make* youth in Sierra Leone. Then, for each aspect, I discuss how ongoing practices of youth were continuous with the recruitment and participation of children in fighting forces.[9]

Child Labor

When Sierra Leone's economy was more agriculturally based, men who could afford it would have large families in order to have a large workforce. This meant marrying several wives, bearing as many children as possible, and including members of the extended family and even outsiders as part of the household. Some of this still remains today: certainly, polygyny is still practiced in Sierra Leone, there is still a cultural bias toward having a large number of children, and the pattern of housing members of extended family and others is deeply embedded in the culture. Children were seen as members of the workforce, and this is true to some extent today (Bøås and Hatløy 2008; Grier 2004).

People are very open about their child labor practices. My palm wine tapper in Goderich told me, "The only reason one has children is so they can work for you. Especially up-country. It is easier to have your children make the farm rather than hire people to do the work. Like now, when I come from work I find my children have cooked and cleaned, and they dish rice for me. At times people can send one child to school, but the rest should be home to work."

Children are a necessary part of any household in Sierra Leone for their labor. I remember taking some time to get used to this phenomenon when I was a Peace Corps Volunteer teacher in the late 1980s. My fellow teachers would stop any passing child to do any type of small errand—"You. Go buy cigarettes for me." Near Rogbom, when I needed directions to a neighboring village, I was told by a stranger, "I'll send a child with you." (With our panic about child abductions, this would never happen in the United States.) Children are expected to take on domestic responsibilities and chores as part of a system that promotes interdependence within household and community relations. In fact, parents who fail to place certain responsibilities on their children can be perceived as neglectful (Bledsoe 1990a).

Labor almost defines good childhood in Sierra Leone: a child who does not work is a bad child. A child might have to sweep the house and compound, get water for the household's morning baths, find wood in the forest and bring it home, maybe go to school, and if not school, then work on the farm or in the garden or help the adults with whatever work they are doing. Children are responsible for driving birds from the fields. A child might have responsibility for caring for a younger child. He or she might be sent on errands. He or she would have to do laundry (by pounding clothes against stones at the river). Children are often the sellers of small items; they wander around with head pans full of onions, bananas, or other produce on their heads. Boys wander around at nightfall with a funnel and a gallon of kerosene, selling it to fill up lamps. Young boys might walk many miles to get five gallons of palm wine to bring back to sell. An urban child would have a slightly different set of tasks but would still be required to work. The child of a fisherman might have the task of caring for the nets.

When I was staying with a Sierra Leonean family in Port Loko for a week, a boy in the house was assigned the task of making sure I had water for bathing in the morning and the evening. I went with him one afternoon to see where he got the water. Port Loko was very dry at that time of year, and water was scarce. He had to climb down a steep narrow path with a bucket, fill the bucket with water from a pit in a valley, and then walk back up the path with the water balanced on his head. I had trouble walking up the path even without a bucket on my head. There is nothing unusual about this in the Sierra Leone context, although I tried to be sparing with my water use after that.

This is not to say that children do not also play, or sometimes try to shirk work. And sometimes, people acknowledge, children are worked too hard (Nieuwenhuys 1996).

Work is an important part of life and learning. Most children can make change and take part in simple cash transactions. Child labor is also learning. As David Lancy puts it, "No one teaches a young girl how to carry water. She will toddle after her older sisters to the stream and, as soon as she is stable enough on her legs, someone will place a small pan on her head and fill it with water and she will carry it home. She has seen it done and, with practice and increasing strength and stature, the pan she can carry will grow larger" (1996, 144).

CHILD LABOR AND RECRUITMENT AND PARTICIPATION

It did not seem unusual to Sierra Leoneans that child labor would be essential to fighting forces. Importantly, there are certain types of work that are primarily children's work, such as fetching water. The rebels needed someone to fetch water, they needed someone to do their laundry. If we think of a traveling band of rebels as a small community, they would need children to allow the community to function. But this is not just the case with rebels. Even the government army needed children to function for the same reasons (Murphy 2003).

The majority of the population of "child soldiers" were children who did standard daily tasks: fetched water, cooked, cleaned, carried things on their heads. And even those children who did more soldierly things—shooting guns, chopping hands—were doing it within a system in which it made sense for children to work alongside adults.

The work of spying was often done by children. Children move around in the process of selling. Children selling can move around almost anywhere; they can pass through public and private spaces, and they are hardly noticed. Also, there were many children manning checkpoints. This fits into the pattern of child labor as well. The adults might have been inside a palm frond shelter, set up to shield them from the sun, drinking palm wine or smoking marijuana, and they would send the young boy to deal with the passing vehicles. Even now, at government checkpoints, it is the young driver's apprentice who most often comes down from the back of a vehicle to pay the ubiquitous bribe to pass.

Thus, within a vernacular in which the labor of children is expected and even required, the use of children as workers in the pursuit of war is not surprising. In interviews with former child soldiers about their time "in the bush" this kind of labor was so unremarkable as to be not worth talking about. Furthermore, when I was interviewing them, safely after their participation in the fighting forces, they were all still routinely doing such labor, even those under the care of child protection agencies.

Child Fosterage

Fostering can be defined as any time a child's primary caregiver is other than the biological parents. The practice of fostering is quite common and is written about extensively in the anthropological literature on family structure in West Africa (see Schildkrout 1973; Goody 1982; Isaac and Conrad 1982; Isiugo-Abanihe 1985; Bledsoe and Isiugo-Abanihe 1989; Bledsoe 1990b, 1990a, 1993; Alber 2003, 2004; Verhoef 2005). Fostering is an umbrella term for a number of different types of relationships. Traditionally, fostering is an exchange; it is not simply that a family takes in a child because the child is in need of care, but that families share the burdens and the rewards of child rearing. The other key point is that fostering is not something that happens only in periods of turmoil; it is an ongoing system.[10]

Fostering is done for a number of reasons. According to the circumstance, a child can be perceived as an asset or a burden: an asset who can work and provide for the home but a burden who costs money for school fees, clothing, and food. Fostering relationships within an extended family are the model for other types of fostering relationships. Among family members, fostering is done to cement family bonds. Fostering creates alliances. For example, if your child is fostered to your aunt, you will be likely to support that aunt in family disputes, or to give that aunt resources. In addition, fostering can spread out the burden of child raising across the family. If a family member is childless, he or she may need a child for household labor and in return will bear the cost of raising a fostered child. In addition, there is a belief that children raised by their own parents are in some way weaker than those fostered and therefore somehow tougher or more socially adept.

I asked my friend M.T., an old buddy and fellow teacher from my Peace Corps days, about child fosterage. He was originally a "village boy" but had been living and working in the city for the last ten years.

Actually, in Krio I phrased the question more like "What does it mean to give someone your child to *mehn* and why does one do it?" If you give your child to your sibling to *mehn* (raise) that draws you closer together, M.T. said. "Originally our people thought that the more children one has the better. More children mean you can make a bigger farm. Also, because the labor of children is indispensable, someone without any children is in trouble, so a sibling might decide to give a childless sibling a child of his or her own. Also it spreads around the economic burden of caring for children to foster them out."

A child may also be fostered to a family member who can help with that child's schooling. If a child of secondary school age lives in a town without a secondary school, he or she may be sent to a relative in a larger town. Although it is normally assumed that parents will be responsible for the cost of formal education, if they cannot afford it, they will sometimes pass on the burden of school fees to a more successful family member. That family member will then expect some minimal labor around the house but, more importantly, will expect that when the child grows up and gets a job, he or she will in turn support other children of the family in need of help. Paying school fees for someone is a kind of cross-generational exchange (Bledsoe 1990c).

Since I made my home base with my friend Wusu and his family near the teachers' college where he taught on the outskirts of Freetown, I was privy to all sorts of negotiations about school fees for his extended family and others. He explained to me, "The expectation is that if someone pays school fees for you, later you will pay school fees for their children, and so on. So an investment in a strange child's schooling is like an investment in your own children's schooling." As Hoffman (2003) put it, in Sierra Leone, children are the unfinished products of social networks. These intergenerational networks continue even in uncertain times.

One day, Wusu's brother-in-law, Henry, was complaining to me about Pa Issa (his wife's and Wusu's uncle). He said, "Pa Issa does not care for his own children; although he has lots of money, sent to him from America, he does not invest it in his children. Rather, he keeps it with a Lebanese merchant in Makeni." Henry was complaining that he should keep the money in a bank, but more importantly, he should help his children (meaning nieces and nephews as well). Since he is old, if his bigger children are trained, then when he dies they can help the younger children. However, since he is not helping them now, some strange person may come along to help them and then they will give help to that person's children instead of to their own brothers and sisters, saying *"papa no bin hep wi"* (father didn't help us). This also adds to the notion of children as resources that require investment. Furthermore, if you do not invest in them, someone else may come along and snatch them away.

Children may also be fostered for training or apprenticeship. If a child takes part in a traditional apprenticeship—whether Koranic schooling, or training in a skill such as carpentry or blacksmithing—the child will usually live with the master as his child. The apprenticeship relationship is slightly different from the usual family fostering arrangement, in that it has a specific purpose and a specific ending time (when the skill has been learned). Also, there is a more formal handing over ceremony in which the parents may pay some small amount as a symbolic payment for the care of the child.

Children may be fostered with a prominent or wealthy person to ensure better opportunities for the future. According to one informant:

> The way our people used to do it before, sometimes the child is small, he does not have sense yet, they take him to somebody. It can be a relative or it can be some prominent person, maybe inside the same village or the same community. What they can normally say is: 'Please, this is your own child, forever. Raise him like your own child.' And as the child grows up, they train him up in such a way to let it get the feeling that this foster parent is his real parent.

This type of fostering was also very common in the first half of the twentieth century when "up-country" children would be fostered to Krio families for their labor. That system is sometimes called "Creole wardship," and it still rankles some non-Krio people. M.T. told me it led to a complicated legacy: provincial children were given to Krios as houseboys—they were mistreated and had to do all the work, but ended up educated and with Krio surnames.

Finally, we come to the type of fostering most common in Western societies, when someone agrees to care for a child who is orphaned or abandoned.[11] In the Sierra Leone context, one's nieces and nephews are considered one's own children, so it is difficult to find a true orphan. This type of fostering is obviously more common after the massive disruption of the war. There are important differences between fostering a child from one's extended family and fostering a "strange" child. A strange child will contribute only to you. You won't have to share the returns. A strange child is more willing to be under your control because there is no one to lobby for him or her. For these reasons, people may even be more willing to take on a strange child than a child of the extended family.

The important point here is the flexibility of family arrangements, and the idea that decisions are made about fostering based on a beneficial distribution of resources, both material and symbolic. Sometimes adults manipulate these relations for their own ends, and sometimes children undo those manipulations. It is also not unusual to have groups of adults and children living together with a range of different family and pseudo-family ties.

FOSTERAGE AND RECRUITMENT AND PARTICIPATION

The language of fostering was sometimes used in RUF abduction. Pa Kamara of Rogbom had two of his children abducted by rebels. In both cases, the rebel who took the child, "asked" for the child. Pa Kamara said at one point the rebels were threatening to cut his hand. They had his hand on the block and the machete in the air. His children were crying and begging for them not to cut their father's hand. That's when one of the rebels saw the ten-year-old boy and said he wanted him. Traumatic abductions often took place according to this pattern. Of course, the parents or guardians were forced, the rebels had guns after all, but it is fascinating that the individual rebel went through the motions of "asking" for the child, holding up the cultural forms of fosterage arrangements. The surprising fact from my interview with the son is that the rebel who wanted him did not just take him on as a soldier, but said he wanted him as his own, to take him home and put him in school.

Once abducted children were away from home, the fosterage tradition made possible a certain ease in creating pseudo-family arrangements. When children talk about their time in the bush, they talk about their commander as a sort of father figure. The commanders' "wives" would direct the labor of the boys, as a foster mother might. In fostering, the bonds between child and carer are not easily broken and are long-lived. A rebel commander and an abducted youth may feel certain bonds exist between them even after demobilization, because their relationship can be understood as a type of fostering. I saw examples of children going to their former commanders to ask for assistance of some kind, even after both had been demobilized and their official relationship was over.

It was unusual, but occasionally families would go to the RUF directly to negotiate for the return of their abducted children. They could draw on the rules of fostering to get children back saying, for example, "You didn't pay the price for this girl." Sometimes they could pay the abductors and reclaim their children. From the Western standpoint, it could be understood as a ransom for a hostage, but in a system where children are routinely exchanged and money sometimes changes hands for their care (for example, in fosterage relationships, parents will sometimes pay a symbolic fee to a master for the care and feeding of a child), these negotiations took place within an already existing system of exchangeable children.

Another important item having to do with fosterage is that we in the West think that children should live with their nuclear family, and they often do. Part of the trauma of child soldiers from our perspective is the experience of being taken away from the biological mother and father. While abduction was often violent and certainly traumatic to families, the particular trauma of being removed from the care of one's biological mother and father does not exist in the same way in Sierra Leone (or is not understood to be traumatic), since so many children are sent to live in other families without their mother or father. To be taken away from home and raised in another family exists as a reality or a possibility in the life of almost every child in Sierra Leone. In fact, living away from one's mother and father can be seen as good for a child, strengthening him or her through emotional hardship but also bringing the child closer to members of the extended family. In other words, in Sierra Leone family is extremely important, but *nuclear* family is not.

Apprenticeship

Here, I must distinguish between learning, which happens everywhere in all sorts of social relations, and education, which I want to think of as an institution, a set of practices and discourses, and a site of strategic deployment and redeployment of cultural and social capital. Education happens in some sort of formal relationship and involves a social transformation. In other words education happens at special times and places where learning is recognized as something that transforms a person. Schooling is a Western institutional form of education, and in Western ideology is the privileged site of learning.[12] In the view of anthropologists Bradley Levinson, Douglas Foley, and Dorothy Holland (1996), schools should also be seen as just one site among many in "the cultural production of the educated person." This terminology will allow us to distinguish between education broadly, schooling as a particular form of education, and learning as omnipresent in social life. There are multiple educational trajectories in Sierra Leone, with schooling the most elite: Koranic school, skills training programs, apprenticeship, secret society training, and so on.

Education is a powerful component in the construction of youth in Sierra Leone and has been, since at least colonial times, a site of political struggles over futures (Shepler 1998; Bledsoe 1992). Even though jobs for the educated are quite scarce, there is still an astounding demand for schooling in Sierra Leone. Aside from formal schooling, apprenticeship is a vital institution for the training of young people into adulthood and often involves fosterage to a master. Scholars of rural West Africa have noted that elders seeking to solidify control over youth try to place tight controls on information they construe as valuable, and protect it through rituals and powerful associations based on secrecy (Bledsoe 1992, 190). In Sierra Leone there is a notion of knowledge (and especially secret knowledge) as power, and gaining that knowledge is seen not simply as filling an empty vessel, but as a powerful transformative experience—not just acquiring knowledge but forging a new identity.

According to anthropologist Caroline Bledsoe, the necessity to work for and compensate teachers forms the backbone of a fundamental cultural theory of child development, aptly summarized by the Sierra Leonean maxim "no success without struggle": "This maxim implies that in order to 'develop' . . . children cannot simply learn knowledge through intensive study: they must earn it (if necessary through tolerating hunger, beatings, and sickness) from those who legitimately possess it, through proper channels of social recompense" (1992, 191).

"[T]he content of knowledge cannot itself bring the rewards of educa-tion, because knowledge does not stand apart from social relations as a detached cultural package. Since blessings legitimate rights to certain domains of knowledge, how children learn—that is, through earning blessings —is as important as what they actually learn" (ibid., 192). And as anthropologist Michael Jackson puts it, "[A]cquisition of luck or benefit depends not on the possession of talent, knowledge, or merit alone, but on earning blessings through a dutiful relationship with a status superior, such as a husband, a teacher, a politician, a business-man—paying him respect, working for him without complaint, serving him faithfully, and doing his bidding" (2004, 171).

APPRENTICESHIP AND RECRUITMENT AND PARTICIPATION

The idea that there is a continuity between educational aspirations before the war and RUF recruitment is not my own. In fact, that is one of the central arguments of Paul Richards's *Fighting for the Rainforest*. According to Richards,

> In a patrimonial polity, where clientelism is a major means through which intergenerational transfers of knowledge and assets are achieved, young people are always on the look out for new sources of patronage. Where they joined the rebels with any degree of enthusiasm it was to see training. The arts of war are better than no arts at all. The army was simply seen as a new form of schooling. Where recruits were gathered together for training in the field, in advance positions, the commander in question would take young volunteers as personal "apprentices," rather than as formal recruits (1996, 24).

He continues,

> For many seized youngsters in the diamond districts functional school-ing had broken down long before the RUF arrived. The rebellion was a chance to resume their education. Captives report being schooled in RUF camps, using fragments and scraps of revolutionary texts for books, and receiving a good basic training in the arts of bush warfare. Many captive children adapt quickly, and exult in new-found skills, and the chance, perhaps for the first time in their lives, to show what they can do.

Stood down boy soldiers in Liberia have spoken longingly of their guns not as weapons of destruction but as being the first piece of modern kit they have ever known how to handle (ibid., 29).

Another piece of this argument, which Richards does not discuss, is that RUF commanders may have wanted a large set of apprentices to enhance their own standing. Some of the former RUF children I interviewed told me that one of their duties was to serve as "bodyguards," or a kind of entourage. In any activity, the more apprentices you have, the more important you are. *Rebel sehf lehk foh bluff* (even rebels like to show off), they told me.

Not only the RUF drew on preexisting models of training. The Sierra Leone Army (SLA) also recruited large numbers of young men in a similar sort of patronage move.[13] This grew out of apprenticeship models as well. A man serving in the army might want to include his son in his daily activities. Given the nature of learning in apprenticeship, it makes sense that young men would participate as "legitimate peripheral participants" (Lave and Wenger 1991). It might mean cooking or cleaning for army men, or sometimes carrying weapons or other equipment. Still, these duties involved young boys in army life. When the RUF and some elements of the SLA merged into the AFRC, some of the army camp followers became indistinguishable from rebel boys and carried out some of the same atrocities.

The CDF is another important example of apprenticeship models leading to child participation in war. Kamajohs, Tamaboros, and Gbethis took in and initiated large numbers of young boys and gave them training and status. As Steven Archibald and Paul Richards explain it,

> Would-be applicants received military training only after initiation as a kamajoi (Mende "expert hunter," lit. "master of marvels"). Such hunters belong to a craft association or guild (Muana 1997). Initiation requires money, or a sponsor to cover the costs. This, it hardly needs pointing out, is rather different from joining an army through the usual routes (volunteering or through conscription). To enroll through initiation into a guild is conducive more to a notion of fighting war as a craft. Warriors tend to see themselves as craftsmen specialists, jealously guarding their individual rights and privileges. CDF fighters became

"professionals" in the sense we might apply that term to a lawyer or doctor in private practice (2002, 355).

Secret Societies

Sierra Leone is well known for its secret societies, which, among other things, play an important role in the transition from childhood to adulthood.[14] According to anthropologist Michael Jackson,

> A child is, to use a phrase of Meyer Fortes's, only an 'incipient person' (1973, 309). Among the Kuranko, only initiation at puberty can create a 'whole' person, a completely socialized adult. Until initiation, children are considered to be 'impure' and incompletely born; physical birth must be complemented by the ceremonial 'birth' undergone during initiation (1989, 76).[15]

In earlier times, young initiates would be "in the bush" with adults for months to learn skills specific to their sex.[16] Adults would be in charge of training youth for several years before they were fully initiated. Initiation into the secret societies ties you to your locale and can be understood as a kind of educational institution. With the increased mobility of people, including youth, there are still ties to specific locales, and tying children to the location of their parents and their tribe is now one of the main functions of secret society initiation. Even before the war, the initiation process had undergone tremendous change from earlier times. As more and more children attended formal Western-style schools, there was less time for them to be sequestered for initiation training. Now training in the bush is usually a matter of days rather than months. Students may come home from school for their initiation during school holidays. In urban places, initiation practices have particularly eroded, and the mix of ethnic groups has led to a syncretism and crisis over what constitutes authentic traditional practices. The war has further shattered these traditions while reconfiguring others.

The secret societies still play an important role in formally marking the boundary between youth and adulthood, in that one cannot be said to be an adult unless one has been initiated. However, since initiation is taking place at younger and younger ages—some girls are initiated as

early as six or seven years old—initiation can be understood as a necessary but not sufficient condition for adulthood. Initiation as a marker
between stages of life has weakened, even though it is still important.

SECRET SOCIETY AND RECRUITMENT AND PARTICIPATION

The CDF recruited child soldiers through secret-society connections.
It was the strength of the society and the affiliated elders that brought
the members. Historian Stephen Ellis notes that CDF-like groups
throughout West Africa have grown out of secret societies. He says,
"Examples of modern militias with a detectable background in traditional initiation societies or procedures include the Mouvement des
forces démocratiques casamançaises (MFDC) in Senegal, the kamajors
in Sierra Leone, the Lofa Defense Force in Liberia, the dozos in Côte
d'Ivoire and the Bakassi Boys in Nigeria" (2003, 4).

Less obvious, perhaps, are some of the ways RUF abduction was like
secret society initiation. First of all, the public performance of initiation
is that children are taken away from the town (the site of social order)
to go live in the bush (the site of powerful forces, both destructive and
generative) to be molded into responsible adults. That is, when the initiation happens, although parents have paid for and prepared for the
event with excitement, mothers cry about the abduction of their child
by the society devil. Similarly, when children were abducted by the RUF,
they were taken to "the bush" for a kind of remaking. In Rogbom, the
children who were abducted by the RUF were called *di wan den we den
bin kehr go* (the ones they took away). That is also how one sometimes
refers to initiates who are in the bush. An old man in Rogbom made
the comparison explicit when he explained (in Temne) that when the
children were taken away, "It's like when the society comes and takes
children to be initiated—with no warning and there's nothing you can
do to stop them."[17] Michael Jackson makes a similar point:

> Indeed the RUF leadership sometimes invoked initiation rites in justify-
> ing its revolutionary method of preparing young boys in bush camps for
> the violent, but necessary, cleansing of corrupt towns under such code
> names as "Operation Pay Yourself" and "Operation No Living Thing."
> For many of the kids who went to the bush and joined the RUF, this
> desire for initiatory rebirth as men of power (purified of the taint of

childhood) may have been stronger than their commitment to the RUF cause (2004, 159).

When initiates come out of the bush, they have a new relationship to their parents. In a disturbing parallel, there are stories of RUF abductees being forced to kill their parents or other family members to distance them from their former selves as civilians and take on the new identity of RUF soldier. Again, Michael Jackson noted the same similarity:

> The abduction of children by the RUF, and their adoption by rebel leaders—who were regarded as fathers, and called Pappy or Pa—recalls the initiatory seizure of children, whose ties with their parents are symbolically severed so that they can be reborn, in the bush, as men. This idea that war—like initiation, or play, or an adventure—is a moment out of time, spatially separated from the moral world, may also explain why many combatants anticipate a remorse-free return to civilian life (2004, 159).

In addition, when initiates are in the bush, they operate under different rules than in their normal lives. They are separate from the community and are fattened up in preparation for the circumcision ordeal by eating great amounts of the richest foods their families can afford. The RUF boys I met talked about a similar kind of plenty in the RUF bush. They told me, "We ate meat every day. Whatever we wanted, we took."

Finally, secret society initiation includes circumcision.[18] Rebel abductees, both children and adults, sometimes had "RUF" or "AFRC" carved onto their chests with razor blades and then had ashes rubbed in the wound to form a scar.[19] This marking is symbolically similar to a circumcision in some ways. People with the scars told me that it was done to mark them irrevocably as members of the rebel factions. In practice, it made people afraid to escape captivity, fearing identification and retribution if anyone found the markings on them.[20]

I am not claiming that abduction by the RUF was identical to secret society initiation. Abduction by the RUF occurred under dire threat, was not seen as beneficial by the family, and was seen by young people as leading to harm. RUF abduction was certainly traumatic for both the child and the family. I am only pointing to certain cultural continuities that would surely resonate with Sierra Leoneans.

How Sierra Leoneans Understand Child Soldiers

The cultural continuities I have just described are so natural as to be unremarkable to most Sierra Leoneans, yet neither they nor I would argue that it was their customs or practices of childhood and youth that led to the worst abuses of child soldiers. The continuities explain, to some extent, why child soldiering took the forms it did in Sierra Leone, but not the existence of child soldiering. How, then, do Sierra Leoneans understand child soldiering?

The rest of this chapter examines some of Sierra Leoneans' own ways of explaining the participation of children in war. First, they see child soldiering as part of an ongoing social breakdown, brought on, in part, by postcolonial economic and state breakdown. Second, they understand "violent youth" as a powerful historical category. I go into some detail about the history of the participation of youth in political struggles from the precolonial period on. Finally, I investigate some Sierra Leonean ideas about the nature of children—simultaneously malleable and unpredictable—that they often call upon to explain child soldiers' worst atrocities.

Social and Economic Breakdown

Probably the number one reason given by Sierra Leoneans for the war is *wi no lehk wisehf* (we don't like/love ourselves). This explanation may sound unbelievable, but a public posting on peezeed.com—a website for Sierra Leoneans in the diaspora—echoes what I heard many times from Sierra Leoneans about what they should do to help rebuild their country:

> There is no sane reason as to why Sierra Leone should rank last in the community of nations in terms of human development while we are greatly endowed with such fabulous wealth and human resources. The first step in rebuilding our beloved country is to genuinely LOVE our fellow Sierra Leoneans and by extension our country. Within and without, we should all help fellow countrymen succeed in whatever ventures they engage in rather than jealous them (sic) or even try to bring them down (Saturday, August 2, 2003).

This reasoning always surprised me, as I thought Sierra Leoneans could perhaps more rightly look outside their country for explanations for poverty. Sierra Leone, like the rest of West Africa, has experienced centuries of unequal exchange with the West—from the slave trade, to extractive colonialism, to the policies of the IMF and World Bank—to blame for its position in the world. There is also the "curse" of diamonds (Smillie, Gberie, and Hazleton 2000; Le Billon 2003) and the war economies that commodities like diamonds make possible (Reno 1997a, 1997b, 1998; Zack-Williams 1999; Nordstrom 2004).

Yet in the simple focus on interpersonal relations lies a critique of the breakdown of a patrimonial political and economic system. What Sierra Leoneans are really saying when they say *"wi no lehk wisehf"* is that people in power do not do enough to help people without power. In particular, the elders do not help the youth, with jobs, education, or even access to corrupt political systems.

Some see recent educational reforms as worsening the crisis of youth. At a "stakeholders conference on education" I heard complaints that the 6-3-3-4 system (six years of primary school, three years of junior secondary school, three years of senior secondary school, and four years of tertiary school), adopted in 1993 (Wang 2007, 34), has caused a lot of the problem. Although it was a response to calls for "Education for All" with a focus on broadening provision of elementary education, it has caused a bottleneck, with large numbers of primary school leavers unable to advance to secondary school. A teacher friend of mine told me that 60 to 70 percent of students fail the exam that would allow them to advance to senior secondary school, and then they become useless in society. "That's when they start hanging out in ghettoes, and all they learn there is how to condemn the system. It just confuses them," he told me. In other words, under this relatively recent education system, there is a specific point when a majority of half-educated youths are excluded from the education system. Krech makes the same point. In his interviews with Ministry of Education Officials, he heard "we at the Ministry of Education in some ways blame ourselves for the war" (2003, 143).

Numerous analysts, both Sierra Leonean and foreign, have sought to understand the war as a "crisis of youth" (Richards 1996; Bangura 1997; Richards 1995; Abdullah et al. 1997; Fanthorpe 2001; Fanthorpe and

Maconachie 2010). Anthropologist and agriculturalist Paul Richards is perhaps the best-known proponent of this theory. Richards and his student Krijn Peters conducted interviews with young rank-and-file combatants from three major factions—the RUF, the AFRC, and the Kamajoh militia. Their analysis shows that in one crucial respect it hardly matters to which faction a combatant belongs: "all tend to share membership in an excluded and educationally-disadvantaged youth underclass. Young combatants are clear about the specific circumstantial reasons they fight against each other. But they are even clearer about what they are fighting *for*—namely, education and jobs" (Richards 1996, 174).

I heard the same thing many times from the young people with whom I spent time. A group of former CDF boys now living in Freetown to attend secondary school gave me a litany of problems of Sierra Leone, mainly, "*di big wan den wicked*" (the elders are wicked). They said that if the RUF had aimed just at the government, and had not started harassing innocent civilians, they would have had a lot of backers. Regarding Paul Richards' thesis that this is a crisis of youth, they said that the greed of the big ones is the real problem. They do not blame their supposed enemies, the rebel fighters, who they see as caught up in similar forces. In fact, they seem angriest at their own big men, who said "come fight for us and we will do everything for you."

Violent Youth Is a Historically Significant Category

There is a long history connecting political violence and youth culture in Sierra Leone. In particular, there is a well-defined identity, shifting in name and shifting in political alliance, but always present. This participation has taken different forms in different eras, at times characterized as young people valiantly resisting oppression, and other times as young people working as the dupes of political elites or involved in violence only for self-enrichment.[21]

The closures at secondary schools in Freetown in the autumn of 2001 are good examples of the role of youth in violent demonstrations. In an effort to decrease tardiness, the principal of Collegiate Secondary School in Freetown had decided to lock the gates of the school in the morning so that latecomers could not enter the school compound. The students protested this policy by blocking traffic outside the school and throwing stones at passing vehicles. A week later, at a different secondary school in Freetown, there was another violent public protest involving the students. Some people who lived near the school had been building shacks on the school land, and the government had been unable to stop them. So, the principal called on the students to go and cause havoc. They rioted, and the school was shut down for a week (basically only punishing the students). There was much comment about these events in the streets and on the radio. These two events, each with a different position with respect to school authority, are important because they illustrate the public assumption that youth would act, even violently, to protest injustice. But there was also much hand wringing by the general public about the state of today's youth, and the fear that even students of relatively prestigious secondary schools are somehow "rebel boys." It did not seem to matter to public opinion the particulars of whether these boys actually fought with the rebels. Any kind of destructive behavior by youth, in postwar Sierra Leone, is termed rebel behavior.

I asked whether people who live near a *poht*—an urban hangout for youth to gather and smoke marijuana—are now more afraid of the young men, knowing they are rebels. "No," they replied, they were always afraid of them. Undisciplined youth has always been a dangerous social class and always controlled for political reasons, for example for thuggery under the APC.

Let us turn for a moment to the historical precursors of this characterization of violent youth (for a more complete elaboration of my historical argument, see Shepler 2010a). Literature from different disciplines on the precolonial, early colonial, late colonial, early independence, and late independence eras establishes that the figure of the young warrior is not new in Sierra Leone.

Regarding the precolonial era, Africanist Vernon Dorjohn claims that war chiefs rarely accompanied their forces into battle, but instead would hire warriors of outstanding ability (*ankurugba* in Temne) to lead their army (Dorjahn 1960). Historians Allen Howard and David Skinner studied the period 1800–1865 in northwestern Sierra Leone and describe the process of war leaders recruiting local boys as part of their networks (Howard and Skinner 1984, 8). Describing precolonial settlements in Sierra Leone, Geographer David J. Siddle notes that "skillful warriors attracted to them bands of mercenaries from surrounding districts ("war boys") who became bandits, terrorizing the areas they controlled" (1968, 50). According to historian LaRay Denzer, the occupation of warrior was clearly institutionalized. Boys were trained specifically for war duties through a system of apprenticeship (Denzer 1971). Kenneth Little, in his exhaustive study of the Mende people, details the changes wrought on Mende society by the encroaching style of warfare starting as early as the sixteenth century. "Within the town lodged the local chieftain and his company of warriors, or 'war-boys', who acted as his bodyguard and private army in the even of a dispute with his neighbors" (1967, 29).

In the colonial era, we see youth mobilized in opposition to the institution of paramount chieftancy imposed by the British. Political scientist Roger Tangri focuses on chiefdom level violence from 1946 to 1956. Citing the Cox report of 1956 (173), he says:

> Bands of "youngmen"—persons other than those holding positions of power in their chiefdom—often counted in hundreds, protested against unpopular paramount chiefs, attacking and burning their property, often alleged to have been acquired illegally. . . . Moreover, although the disorders involved large numbers of "youngmen," they were not popular rural revolts against the elders. There was widespread protest against the general mal-administration of those in power, but . . . the violence was often instigated and guided by elders belonging to opposition "ruling" houses, who sought to have the incumbents ousted from their positions of authority in the chiefdom, and then to supplant them with their own nominees (1976, 313).

Tangri further explains that the chiefdom level riots of the mid-fifties were based on, "A symbiotic relationship . . . between opponents of the local establishment, who wanted to further their own interests, and discontented 'youngmen,' who demanded an end to the abuse of power by the ruling elite" (317).

This pattern of recruitment of young men for political violence continued into the independence era. From the discipline of art history, in a description of the urban masquerade societies (perhaps best understood as the Krio version of secret societies, drawing, however, on Yoruba traditions) of Freetown in the 1970s, John W. Nunley also discusses the political culture at the time. He notes that the early APC organizers recruited young men from the Firestone and Rainbow Odelay societies[22] as thugs, used to rig elections and threaten voters (1987, 59).

One can conclude that the phenomenon of groups of violent young men in the service of opposition political projects has occurred at least throughout the last two hundred years and became more intense in moments of political uncertainty, of which there have been many. Clearly, there is a continuity in the practice of recruiting disaffected youth into violent political protest, and a tradition of youth violence as an expression of wider political discontent, and when Sierra Leoneans talk about "those rebel boys," they do so in ways that reflect that continuity.

I am not the only one to have noted the tradition of youth violence as an expression of wider political discontent; but I saw in my fieldwork that in the recent conflict, local level violence often played out along preexisting lines of lineage versus lineage, and that in any village it was often only the property of the local elites that was completely destroyed. When violence came to an area in the form of RUF rebels, they often recruited local youths into their violent program by invoking preexisting models of violence against local elites.[23]

Ideas about the Nature of Youth

We must also look to Sierra Leoneans' ideas about the nature of youth to see how they explained child soldiering. Some old Sierra Leonean friends of mine discussed the common wisdom on RUF recruitment of children:

> If an older person went to go join the rebels (perhaps seeing all the loot they were getting) the rebels would feel his chin to see if he had any beard. If so, they would send him away saying "*we no want you Papay*" [we don't want you old man]. The rebels only want young boys and girls because they are more easily controlled. If you tell them to kill they will. A big man "*no get da maind de*" [isn't brave enough]. "*Pikin no get waif, he no get pikin den. Rebel den no de frehd dai*" [A child doesn't have a wife, he doesn't have children. Rebels can't be afraid to die].

On one hand, children are understood to be easily controllable and not afraid of death. This is not a consideration only for rebels. I interviewed

Obia, a local commander of the Gbethi—the ethnically Temne branch of the CDF—at some length about his group's decision to use child soldiers. All of the CDF forces had certain laws that could not be broken, or a fighter would lose his magical powers. Obia told me that the young boys found it easier to keep to the laws of the society. He gave many reasons (some of which I will discuss further in chapter 5 when I discuss the CDF in greater detail), but here I want to focus on his comments regarding the malleability of children.

o: Then the laws now, the very small boys, if they're small until about ten or fifteen, he will have understood all the rules. So, he won't damage the laws. A matured person, if you join him today, the law that you give him, he won't be able to carry that law for long. That's why even some of the people die at the warfront. But when he's fully matured . . . we have things that you shouldn't do. We have what you should do. So when they give them those laws, they go behind that [they break the laws secretly]. So when they go to the warfront, they die. If they had gone along with the law, they wouldn't have died.

ss: So the child is more able to keep the laws.

o: Yes.

ss: More than the big ones.

o: Yes.

ss: Why? Because he's not used to . . .

o: For one, most of the important laws that we have, the things you are not supposed to do at all . . . the woman, the woman who you've not married, it's not right for you to follow her. OK, that's another problem. However you wash, that *yanaba* will still be on you, for forty days. So it's not right to do it. So, a child who doesn't do it . . .

ss: Mmm.

o: So that's one point. We have some foods you're not supposed to eat. Like nut oil for example [the cheaper darker oil from inside the palm kernel], you're not supposed to eat it. Pumpkin, you're not supposed to eat it. So, a small child will be able to control himself, but a big man, he's not able. So if you enjoy that meal, if, like, the time when they attack, if you enjoy it today or yesterday, today they attack and they say everyone go there, when you see them follow, they go and stay there [they die]. . . . So, that's our problem. . . . But a small child,

that law, he'll be able to do it. Because for a small boy, even to have a
girlfriend is not easy.

On one hand, children are easily controllable and because they are not as
easily tempted, may find it easier to control themselves; but on the other
hand, children are understood as being capable of inhuman acts. Childhood is a dangerous time since children are understood to be not yet
fully human. These theories of the nature of childhood might seem like a
contradiction, but as anthropologist Mariane Ferme points out, "Mende
representations of childhood are fraught with ambivalence. Given that
power is inscribed within an order of concealment, people who are most
manifestly devoid of it, like children, might in fact conceal it in unexpected ways" (2001b, 197). She continues, "[I]t is precisely when children
are regarded as insignificant—as liminal beings between the world of ani-
mality and madness—that they are perceived as most dangerous" (198).

These ideas were echoed by an old friend of mine from my days as a
Peace Corps teacher. Sonny Joe was the vice principal of the secondary
school where I had taught, and he had seen some tough times during
the war. He had found a teaching job at the technical institute in Bo,
in the south, but he lived alone, cut off from family, with just enough
money to keep himself fed and housed. Whenever I found myself in Bo,
I would always stop by and we would drink a few beers for old times'
sake. Naturally, we would discuss my research findings over those beers.
One evening he told me, "The young ones, 'na den danger' [they're the
most dangerous]. In the RUF they performed the worst atrocities. In
the CDF, sometimes the young ones are the most powerful witches. A
lot of the CDF power comes from witch and sometimes young people
are even stronger witches than old people." He repeated the notion that
young fighters were more ruthless because they had no wife or children
to worry about.

In particular, in Masakane I heard about child soldiers as young as
three, called "bao tchie" in Temne, who were brought into the society
precisely because of the strength of their magical powers. This is a kind
of child soldier we do not often think about, and it is these kinds of conditions of childhood (for example, strength of magical power) that we
do not take into account when trying to explain why factions chose to
use child soldiers.

I noted earlier in this chapter that one of the common reasons cited by child protection NGOs for the use of child soldiers is their malleability and weakness. Sierra Leoneans have a more contradictory theory of childhood that sees children as liminal and unformed, and therefore more capable than adults of inhuman behavior.

What Sierra Leoneans Find Horrifying about Children in War

Although there are continuities, and in some ways child soldiering made sense within the Sierra Leonean vernacular understanding of childhood and youth, this does not mean that Sierra Leoneans were not dismayed by the phenomenon. What, then, was horrifying to Sierra Leoneans about the participation of children in war?

In essence, the activities of children and adult combatants were not that different (though children did much more domestic labor and perhaps children performed the worst acts more easily). When Sierra Leoneans talk about the experience of facing child soldiers as the civilian targets of violence, in addition to the horrors they faced, they point to the added impact of facing an inversion of hierarchies. "The one who did this to me was just a little boy!" or "*A abul bohn am*" (I am old enough to be his parent). The *fityai* (disrespect) involved was literally adding insult to injury.

Sierra Leoneans also worry about the long-term impacts of the war on child soldiers, and the idea that as they grow older, those troublesome boys will become troublesome men. This set of children has "bad training," and may not be salvageable. This conclusion rests on another assumption about the nature of children, that they must be properly trained in order to mature properly. What is disturbing is not a lost innocence (as in Western discourse) but a separation from family and training, and the idea that the nation faces the loss of a generation.

Another concern is that improper initiations into secret societies may bring bad supernatural effects, or a breakdown in the power of secret societies. Although the societies have always adapted and responded to historical circumstances, the war has had an undeniable impact. The war disrupted the initiation activities of societies, and this led to various crises. There were populations of girls who could not join the secret society because the society often could not organize its ceremonies in the midst of war. In addition, the RUF rebels sometimes targeted the sacred spaces of the secret societies in a move against the sites of traditional power. There are reports of RUF commanders abducting society leaders and making the initiation part of the moving rebel community. A telling sign of the power of the women's society was that even rebels were afraid of having sex with (that is, raping) uninitiated girls. In the postwar period, there is the problem of reinitiating girls who may have been improperly initiated the first time.

Conclusion

This chapter has touched on the social, cultural, and historical factors that help explain the use of child soldiers in Sierra Leone.[24] The focus on social, cultural, and historical continuity does not change the fact that the war was an extraordinary event, and a horrible experience for almost everyone involved. Sierra Leoneans will be recovering from the experience of war for decades to come.

To understand childhood and youth in Sierra Leone, one could start with UNICEF data. It paints a picture of a childhood of deprivation, always in distinction to the "ideal" Western childhood. Or we could romanticize it in a kind of Rousseau-like child-as-noble-savage move. The theoretical danger is an extreme cultural relativism that approves any "traditional" practice for the sake of its traditionalness. This can be just as insidious as a fanatical devotion to a universal definition of childhood that always finds African childhoods wanting. The real challenge is to understand Sierra Leoneans' different model of childhood, which works in its own cultural milieu, without condemning or valorizing. The chapters that follow address the challenges to the Sierra Leonean model of youth brought about by the war and the postwar.

2

Child Protection Deployed

The Bo Interim Care Centre

One Sunday morning in late March 2000, I took the government bus to
Bo, the capital of Sierra Leone's Southern Province, and found someone
to give me directions to the interim care center (ICC) for former child
soldiers. I knew the center was cosponsored by Christian Brothers (a
local NGO) and the International Rescue Committee (IRC, an international NGO) and that it was located on the eastern edge of town. Late
March is the very beginning of the rainy season, and big black clouds
were threatening. The breeze was a welcome change from the heat of
the previous months. When I arrived at the ICC, I was told the person
in charge had gone to church and that I should come back in the afternoon. When I came back in the afternoon, he still was not around, but
one of the caretakers, Grace, agreed that I could hang around and wait
for him. I recognized a few of the children from another ICC, and even
more recognized me (at this point I had already spent six months at
another ICC near Freetown and had visited several others briefly). This
reassured Grace, and we all played bingo for a while.

Ibrahim, another of the caretakers, came along and joined the
game. There was also draughts (known as checkers in the United
States), snakes and ladders, ludo (something like parcheesi), *tehtehbol*

(tetherball), and football (soccer). There were about twenty-five children at the center, though Grace told me they were expecting up to two hundred at a time. There were a few adults scattered around. Everyone but me was Sierra Leonean.

The center was located on the edge of town. There was a large main building that used to be a residence and a large mango tree in the yard that people gathered under for shade. Behind the house were a block of latrines and a large kitchen where several women prepared food for everyone every day in two giant cauldrons (one for rice, one for sauce) over an open fire. Inside the house were wooden chairs and benches and chalkboards that made some of the rooms look like classrooms. In another room in the back were mats for the children to spread on the ground at night for sleeping. On the walls of the main room were some pictures from Bible stories and a timetable of activities.

At the time of my visit, there was only one girl at the center, Rugie. She told me she was from Kabala (in the north) originally, but was raised in Kono (in the east.) I wondered why she was in the ICC that served the south, but decided not to push her. She seemed pretty mixed up. She told me that when she was in the bush she went to Maskita's father's village (Sam Bockarie, or Maskita, was a well-known and much-feared RUF commander.)

That day I met a boy everyone called Political. He was around twelve years old. He attached himself to me immediately and had a lot of questions for me about America, and a lot of jokes about his situation. The staff at the ICC laughed and told me that Political (his real name was Peter) was always attracted to the white folks who came around. They also told me they had tried to reunify him with some family in Bo town, but he had refused to stay with them. After my day at the ICC, Peter asked to walk me back to my hostel. He was quick and funny and I enjoyed his company.

I went back Monday morning to spend the whole day at the ICC. I finally met with Brother Alex, the head of the center. He explained to me how Christian Brothers and IRC split the responsibility for running the ICC. Everyone gathered for morning assembly, exactly as Sierra Leonean school children do. We sang Christian songs and the national anthem (luckily I could sing along thanks to my years as a teacher in Sierra Leone). There was an inspirational talk about religious tolerance

("In a way we are all the children of God. Even to argue about religion is a sin.") The children were standing in lines, trying to keep quiet, punching each other on the arms and laughing to themselves.

Weekdays had a strict timetable at the ICC. The morning was taken up with school-like activities. Instead of teachers, they had "animators"[1] and instead of Class I, Class II, Class III, and so on they had Group I, Group II, and Group III. As a former math teacher, I decided to sit in on Group I math lessons for a while. It was dismal. The woman leading the class copied an exercise incorrectly onto the board and then asked the children to perform a task that was made impossible by her copying error. (The math teacher in me could not stand it and I had to show her the mistake. She corrected it, but in doing so only further confused the children.) It seemed to me that the main lesson was how to sit still and keep exercise books, although the classes were more informal than regular school.

Next I went to Group III. They were learning the Bible story of Joseph. The animator said, "You see the father and the son are crying. Why?" One of the boys responded, "Because they haven't seen each other in a long time and they didn't know if the other was alive or dead." The animator said, "Some of you, the same thing will happen when you go back to your families, no? Remember when Mariama's father came to take her? They both cried, not so?" After a while, the children were bored so he gave them some math problems to do.

Lunchtime! As a special visitor I was called to eat away from everyone else. The children each got a colored plastic plate with "combat" (a combination of rice and bulgur) and some cassava leaf sauce. There was some grumbling from the children about the bulgur since they thought they should be given rice only, and some of the new boys refused to eat it. (Bulgur is aid food from the United States and not part of the normal Sierra Leonean diet). I said I *liked* the bulgur for a change and they just laughed, saying, "Well, they should keep it in America for the people who like it."

In the afternoon, there was a visitor from the Planned Parenthood Association of Sierra Leone (PPASL), a Sierra Leonean, to talk to the children about "well body business" or health. He started with family planning. "What is a family?" he asked. He wrote, "Father, Mother, Child" on the board in English but spoke only Krio. He said there are five steps to becoming a big man or woman: (1) learn book OR learn your work; (2) secure a good job; (3) find a faithful partner; (4) get married; (5) have children. He said it is not good to skip steps or to have sex before your time. He told scary stories about STDs and what can happen to a girl who gets pregnant too early. Overall, the message was do not have sex until you are older. All the boys were teasing one of the older boys, Sheku, that he had already started having sex. He was embarrassed, but he admitted it.

After the Planned Parenthood presentation, we all went under the mango tree for "Reintegration Plans." One of the staff went around a circle of the children and asked each child what he would do when, after reunification, the chief told him he had to take part in communal labor. Everyone answered correctly, "I would make myself humble and do whatever the chief asked me to do." The staff member personalized the question for each child, something like, "But you, you are always fighting, will you continue those bad habits?" or "Remember yesterday when we asked you to get wood and you refused? Is that how you will behave in the village?" When the staff member got to Peter, he personalized the message this way: "You, Peter, you love the white man. You always say that anything black men are involved in is rotten. What will you do when the chief asks you to do communal labor?" Everyone, including Peter, laughed at this version of things, but Peter also knew the right way to answer.

According to the timetable, next we were supposed to do "Arts and Crafts." The strict timetable seemed to fall apart as the day wore on and only a few seemed to have the energy to participate. Auntie Susan was the Arts and Crafts person, and she taught the boys how to make "chop covers" (small blankets made out of yarn to put over covered serving dishes to keep food warm) by unraveling sweaters and using the yarn on a simple wooden loom. I am sure whoever donated the sweaters never thought they would end up being used in this way. Some of the boys were really good at it and enjoyed the activity. Many other boys were in the sleeping area or playing other games. The adults sat around under the mango tree and settled small disputes among the children for the rest of the afternoon.

They also had a copy of the employment guide from the IRC (the International NGO who employed them) they were complaining about.

In some ways this was a pretty good job for postwar Sierra Leone. Many of them had worked as teachers before the war, but the Ministry of Education had not paid teachers for a long time.

Rugie came running to the adults, complaining that a new boy who had arrived at the center unaccompanied with nothing but his demobilization form was wandering from room to room checking everything out in a suspicious way. She suspected him of being a Kamajoh spy. She still had the nervousness of the bush about her. When the staff tried to reassure her, she started bragging to the assembled boys that she knew how to fire all the guns, even an RPG (rocket-propelled grenade launcher). One small boy said, "That's a lie!" She countered that RPGs were generally fired by small boys so why shouldn't a girl fire one. Then everyone got into the argument, saying, "Everyone knows that only big men can fire RPGs because of the backflash," and basically ridiculing Rugie's claims as a way to show their own expertise. Even the adult staff got into the argument. Rugie said, "If you don't believe me, bring one and let me test it on you." I thought this bragging to try to get credibility as a real fighter was especially interesting juxtaposed with the recitations of the Reintegration Plans exercise earlier that afternoon.

I continued going to the ICC every day and got to be an accepted part of the community after a while. After several days, some new "bad" boys were brought to the center. They refused to obey directions, and left the compound to find the nearest *poht* where they could smoke "stuff" (marijuana). Apparently, Political had taken up marijuana smoking while "in the bush" but had stopped recently. The new boys were a bad influence on him, the animators told me. The next day, we all gathered early in the morning to go to Kenema[2] for a football match against the boys of the Kenema ICC. The ICC had hired two *poda poda* (minivans) to take us all to Kenema, and there was a lot of excitement. However, Peter and the other bad boys were nowhere to be found. It was finally discovered that they had all decided to leave for Freetown and the excitement of the big city. They left with the clothes on their backs and their ICC-issued sleeping mats to sell to pay for transportation. This really put a damper on the plans for the football match, but we left anyway.

*   *   *

What I have described above is a slice of the complex social practice at one interim care center, with similarities to and differences from the several other ICCs I visited. I am definitely not claiming that the ICCs were not doing their job; indeed, I would argue that multiple, overlapping spheres of social practice exist in any institution, and on-the-ground practice is sometimes developed in contrast to official ideology. In ICCs, individuals are shifting identities in relation to each other and the institution in which they are all located. Staff at one moment may be saying the words they know they are supposed to say, and in the next laughing at one of their charges, and in the next complaining about the conditions of employment. Children shift from rehearsing "appropriate speech," to grumbling about the food, to bragging about their accomplishments as soldiers.

My goal in this chapter and the next is to reveal the many layers of social practice in interim care centers. The preceding chapter discussed Sierra Leoneans' own concerns about child soldiers, and some of the cultural underpinnings of youth in Sierra Leone that inform those concerns. Here the focus is on Western interventions for child soldiers: the NGO-based definition of who and what is a "child soldier." It explains where the interventions come from ("lessons learned" from earlier child soldier projects, drawing on colonial history and development discourse, as well as on child development in the fields of psychology and education). I then move on to a description of the child protection system in Sierra Leone (government, international NGOs, local NGOs) and my interpretation of the theory behind their interventions.[3]

Transnational Interventions in Postwar Sierra Leone

International aid in postwar Sierra Leone takes many forms. In the immediate postwar period, the United Nations had a large presence with a large peacekeeping force; so did other UN-sponsored agencies such as the UNHCR (UN High Commissioner for Refugees), UNDP (UN Development Programme), UNICEF (United Nations Children's Fund), WFP (UN World Food Programme), and others.[4] Multilateral and bilateral aid programs from Europe, North America, and Asia administered projects in health, agriculture, education, peace building, and many other development activities. In addition, there were internationally sponsored justice initiatives, such as the Truth and Reconciliation Commission (similar to the famous South African Truth and Reconciliation Commission) and the Special Court for Sierra Leone (a hybrid court slightly different from the International Criminal Tribunals for Rwanda and the former Yugoslavia).[5]

Special Interventions for Child Soldiers

Interest in child soldiers grew out of international work on child rights generally, and on the heels of other target groups such as child laborers, street children, and "children in especially difficult circumstances." Indeed, historian Dominique Marshall traces humanitarian sympathy for children in times of war back to Herbert Hoover's World War I relief activities and the foundation of the Save the Children Fund in London in 1919 (Marshall 2002). Although UNICEF and various child protection organizations had been working on the issue of child soldiers since the 1980s, the issue really came to the fore in the mid-1990s. The turning point in the modern history of the issue was the publication of Graça Machel's groundbreaking work on for the United Nations entitled *The Impact of Armed Conflict on Children* (1996). Sierra Leone has a special role in the history of the growth of the issue of child soldiers: the 1999 Lomé peace accords between the Government of Sierra Leone and the RUF is the first time (doubtless at the behest of the international brokers) that child soldiers are explicitly mentioned in an international peace agreement:

Lomé Peace Agreement

Article XXX: Child Combatants

The Government shall accord particular attention to the issue of child soldiers. It shall, accordingly, mobilize resources, both within the country and from the International Community, and especially through the Office of the UN Special Representative for Children in Armed Conflict, UNICEF and other agencies, to address the special needs of these children in the existing disarmament, demobilization and reintegration processes (Peace Agreement Between the Government of Sierra Leone and the Revolutionary United Front of Sierra Leone, Lomé, Togo, 1999).

Since the 1999 Lomé peace accords, a number of governmental and nongovernmental institutions have come into being in Sierra Leone to address the issue of war-affected children. Interestingly, even the United

Nations peacekeeping force in Sierra Leone, UNAMSIL, employed a child protection officer. This was the first time that such a position was funded within a peacekeeping mission (Shepler 2010b). Less surprisingly, UNICEF, and to a lesser extent the Government of Sierra Leone's Ministry of Social Welfare, Gender, and Children's Affairs (MSWGCA, henceforth simply the Ministry), were the biggest forces in the postwar child protection institutional landscape. They oversaw international and local child protection NGOs, including Save the Children, Christian Children's Fund, War Child, Defense of Children International, and many more. In what Sierra Leonean scholar Ibrahim Abdullah has characterized as a "donor driven economy," child protection was a growth industry in postwar Sierra Leone.[6]

DRAWING ON "LESSONS LEARNED" ELSEWHERE

Of course, children had been involved in war in Sierra Leone before the huge influx of NGOs and the growth of international coalitions and standards. As early as May 31, 1993, the Government of Sierra Leone announced and delivered to UNICEF 370 child soldiers from its armed forces—including ten girls—who were absorbed into makeshift demobilization centers prior to planned family reunification. The outcome for these children at the end of the year was an early warning about the difficulty of designing effective programs: although 43 percent had been reunified with their families, 27 percent were stuck in "transit" care and the remaining 26 percent had dropped out of the program as a result of unfulfilled expectations (Brooks 2005). This early failure taught policy makers in Sierra Leone to look elsewhere for successful models of reintegration programming. As UNICEF Sierra Leone reintegration officer Andy Brooks puts it:

> Experiences from Liberia and Angola showed that if demobilization consists of little more than registration, distribution of nominal benefits and discharge it could be disastrous. In both countries children were accelerated to their communities at a pace that outstripped any available follow up and high numbers were re-recruited (Brooks 2005).

Therefore, when programs were designed in the aftermath of the 1997 junta period, with the SLPP back in power and the ground once again safe for NGOs, many models for dealing with a population of child soldiers were imported from earlier conflicts in other countries, particularly from UNICEF "lessons learned" documents (David-Toweh 1998; Legrand 1997, 1999). Programs in postwar Mozambique, Angola, Uganda, and Liberia were most often called upon as examples, since those are the nations in Africa that had experienced conflicts since the advent of international concern with child soldiers (Green and Honwana 1999; Honwana 1997, 1999; Ehrenreich 1998; Wessells and Monteiro 2000). The point is that a great deal of the language and the institutions were imported into Sierra Leone from other postwar nations around the world, including Latin America and Asia. This is not only true in the case of child soldiers: the Truth and Reconciliation model was imported wholesale from elsewhere as well, and the Special Court for Sierra Leone is based the International Criminal Tribunals for Rwanda and Yugoslavia. This expert knowledge was usually welcomed by Sierra Leoneans dealing for the first time in living memory with the aftermath of war, but as I noted in this book's introduction, it also represents a kind of abstraction of the experience of war and a homogenization of the experiences of child soldiers around the world.

THE CHILD PROTECTION SYSTEM FOR CHILD SOLDIERS

Child protection activities, including disarmament, demobilization, and reintegration (DDR) programming, carried out by international organizations are generally "rights-based." That is, they start from the assumptions of the Convention on the Rights of the Child and other key child rights documents. Despite high-level discussions within the UN about the need to link reintegration and justice in order to assure community acceptance, the key assumption behind the entire "child soldier" and DDR system ended up being that child soldiers are innocent victims, that as children they cannot be held responsible for their actions. The assumption is that children are by definition innocent and malleable and that any violent actions in which they took part must have been caused by adults (compare this to the Sierra Leonean model of childhood presented in chapter 1). We can think of the goal of humanitarian programs to "return stolen childhoods" through a machine that puts child soldiers in one end and produces innocent children at the other.

In Sierra Leone, this is how that machine was designed: the following activities were undertaken by UN peacekeepers and child protection agencies in Sierra Leone, and make up the set of modern techniques for remaking child soldiers.

1.   Disarmament and demobilization[7]
2.   Interim care
3.   Psychosocial programming
4.   Schooling and skills training
5.   Family Tracing and Reunification (FTR)
6.   Alternative care (for some)
7.   Follow-up visits and community support[8]

Background of the Interim Care System

The purpose of an interim care center (ICC) is reflected in the name; it is meant to be a place where children are cared for in the interim between demobilization and reintegration, a place where children can receive medical treatment and regroup in a safe and secure environment. During the period of my fieldwork, there were several ICCs in operation around the country, at least one for each region. After demobilization, a child was sent to the ICC in his or her region of origin, or in the region in which he or she hoped to be reunified with family. For example, a boy may have come originally from Kono in the east, but since Kono was still under RUF control, UNICEF would not support reunifying him there. Or, he might not know whether any of his family was still living in that area, So child protection workers would interview the boy and try to find other family he might be sent to live with. If he had some relatives in Freetown, he might be sent to the Freetown ICC while his family was traced.

ICCs were operated by different child protection NGOs in different regions, and located in the major towns. The location and sponsorship of ICCs shifted even during the course of my fieldwork. Usually NGOs with an already existing reputation in a given region were chosen to administer ICCs in that region. Some ICCs were started out of already existing child protection programs. For example, the Christian Brothers program in Bo in the south built on an already existing street children program. In each region, an NGO was identified that could take on the project, and UNICEF then gave funding and support. Some ICCs were

Table 2.1. Locations and Sponsors of Main Interim Care Centers

Location of ICC	Region served	Sponsoring NGO(s)
Lakka	Freetown, Western Area	Family Homes Movement (local, Catholic) COOPI (Italian)
Lungi	North	Caritas Makeni (local, Catholic)
Port Loko	North	Caritas Makeni
Makeni	North	Caritas Makeni
Bo	South	Christian Brothers (local, Catholic) IRC (international)
Kenema	East	KDDO (later Caritas Kenema) IRC (international)
Daru	East	Save the Children UK (international)

run entirely by local organizations, some were run by international NGOs. Often, there was some kind of joint sponsorship, with the local NGO providing personnel and the international NGO providing materials and expertise. The number and sponsors of ICCs were changing all the time during my fieldwork. Some small programs were trying to become ICCs, and some international NGOs were taking over from local NGOs. The biggest players in each province are presented in table 2.1.

Over the course of the war, there was little child protection activity during the NPRC era (the early 1990s), but there was a small increase in the mid-'90s (between the NPRC era and the 1997 junta). The actors then were UNICEF and some small local NGOs, usually affiliated in some way with the Catholic Church (for example, Children Associated with War, and Christian Brothers). After the intervention, starting in 1998, there was a slow and steady increase in child protection NGO activity. UNICEF and the Ministry developed nationwide programs based in the small existing programs run by local NGOs. Over time, more international NGOs got involved and started taking over from the small local NGOs (for example, IRC took over many areas of activity from Christian Brothers). During the course of my fieldwork, I spent several intensive months visiting the ICC in Lakka almost every day (Lakka is the main site of analysis in the next chapter). I also visited ICCs in Lungi, Port Loko, Bo, and Kenema for periods of three days

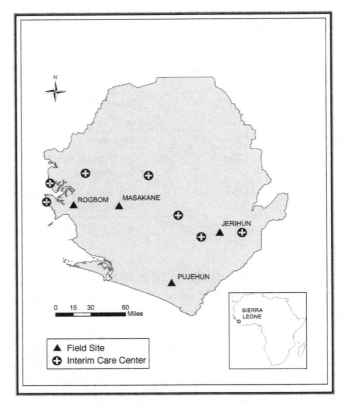

Figure 2.1. Approximate field sites and interim care centers (ICCs).

to a week.[9] Each ICC had an operational area to cover, that is, children from a particular province—West, North, East, or South—were sent to the ICC in that province.[10]

The population of an ICC was by its very nature variable. Demobilized children were brought in groups, and they left one by one as their families were found and their reunifications negotiated. Naturally, some cases took longer than others. The Ministry Guidelines, developed along with UNICEF, called for a period of six weeks in interim care, though this guideline was rarely met, with some children staying at the interim care stage for years (Williamson 2006). During the course of my fieldwork, I saw ICCs operating with as few as ten children and as many as three hundred. Despite estimates that girls and boys were abducted by the RUF in roughly equal numbers, in every ICC there were a few girls, but never more than a handful compared to the number of boys. Of child ex-combatants, children in ICCs were almost always RUF or AFRC. CDF children were not mixed in among this population because most of them had not been separated from their families. Not only child ex-combatants were housed at ICCs. "Separated Children" were also taken to ICCs. These were children who had somehow become separated from their families during the war and could also benefit from family tracing and reunification.

ICCs were designed to provide for a number of needs of recently demobilized or separated children. They provided meals, clothes, and rudimentary supplies. At national Child Protection Committee meetings including UNICEF, the Ministry, and representatives of all the major child protection NGOs, there were struggles over defining the appropriate standard of care in ICCs: it should be enough to meet agreed upon needs, but not too much that children would find it difficult to reintegrate into their original communities (Brooks 2005, 21). Each ICC had a staff of trained Sierra Leoneans with various backgrounds who worked with the children to "bring them back to normal." Some of the staff were young men, some were older women. In a secure environment, they provided medical screening and treatment (many child ex-combatants came into the centers with sexually transmitted diseases, for example), psychosocial activities, some schooling or skills training, some counseling about what would happen upon their return to their villages, and family tracing and reunification programs. Each ICC did things a little differently. For example, the quality of food and entertainment varied across institutions, as did the level of educational support. These differences were crucial for the children in the centers.

Psychosocial Programming

Victims of war often exhibit symptoms of what is called post-traumatic stress disorder or PTSD. According to the Western psychiatric definition, symptoms of PTSD and related stress reactions common in children include avoidance/numbing, as in cutting off of feelings and avoidance of situations that provide reminders of traumatic events; insomnia, inability to concentrate, and intrusive reexperiencing, such as nightmares and flashbacks; lethargy, confusion, fear, aggressive behavior, social isolation, and hopelessness in relation to the future, and hyperarousal as evidenced in hypervigilance and exaggerated startle responses (*Diagnostic and Statistical Manual of Mental Disorders [DSM-IV]* 1994, 427).

The notion of PTSD has come under attack from many quarters, but at the time of my fieldwork it was still the primary model for dealing with children affected by war.[11] Healing children after war, within a Western framework, has come to be associated with particular psychosocial symptoms of "trauma," and particular psychosocial remedies, largely centered on the individual child. When I tell people in the United States about my research, they often respond, "Oh those poor children, they must need therapy!" This reaction is in line with a psychologistic hegemony in the West that understands trauma as something that happens to an individual and, furthermore, something that can be cured in an individual. (Imagine, for example, what might have been prescribed for the survivors of the Columbine High School shootings: psychologists would be on hand to deliver therapy and opportunities to "process" the trauma.)[12] I heard many times from Western NGO personnel in Sierra Leone, "Can you imagine the size of the problem? In the whole country there is only *one* psychiatrist!"[13]

With adults, treatment for PTSD involves "cognitive restructuring" and is usually performed through talking about the trauma on numerous occasions. Sociologist Chris Gilligan (2009, 119) notes that since the end of the Cold War, humanitarian interventions to provide psychological assistance to children exposed to political violence have become commonplace since there is a widespread conception that children exposed to political violence are highly vulnerable to psychological trauma. With children, drawing, painting, and storytelling are often used with the aim that trauma should be relived. Though recovery programs encourage the retelling of trauma to process it in some way, young former combatants in Sierra Leone, as reported in Peters and Richards (1998) said that the two things they most want are education (vocational training and skills especially) and to forget the war. This is in contrast to the usual treatment for PTSD which requires a remembering, a recitation of horrors, to deal with trauma.

For children, what is often prescribed is art therapy, (See figure 2.2, a picture drawn by a former child soldier in an ICC.) For example, at

Figure 2.2. Drawn by a former child soldier.

the time of my fieldwork the only activity of War Child Sierra Leone, a Dutch NGO, was an art therapy program for war-affected youth in Sierra Leone. Emilie Medeiros, a clinical psychologist who worked with Handicap International in Sierra Leone from 2003 to 2005, believes that mental health was neglected in programming for child soldiers and needs to be incorporated further into designing policy, training, and interventions (Medeiros 2007; see also Clifton-Everest 2005); she agrees, however, that there is room for more culturally appropriate frameworks. This approach has become so widespread that it now appears as "common sense" to many people living in Western societies. I believe this orientation toward self and the associated assumptions about where the meaning of reality is located have their origins in specific developments in Western thought and culture.

The realization that the social aspects of trauma can be as important as the individual aspects has led social concerns to be included in treatment. Even the most psychological analyses, for example, that of psychiatrists Peter Jensen and Jon Shaw, make this point clear:

[A]lthough war is undoubtedly "stressful" for children, the concept of PTSD (as usually employed) may have limited applicability to the full understanding of the effects of war on children. War usually represents a chronic, enduring condition, in which the entire context and social fabric may be dramatically altered. Entire nations and cultures may be disrupted, whereas most events leading to PTSD occur under much more limited circumstances (1993, 698).

"Psychosocial" evolved as a compromise term between the view that trauma is in individual heads ("psycho") and that trauma is something that happens to communities ("social"). Lindsay Stark, Neil Boothby, and Alastair Agar at the Program on Forced Migration and Health at Columbia University explain that "the term 'psychosocial' encompasses social, cultural and psychological influences on well-being" (2009, 526). Psychologist Michael Wessells avers that it is a composite term that includes individual effects (not only trauma, which has been emphasized in the field, but also depression, anxiety, and somatic disorders) and social effects felt at family, community, and wider levels. The psychosocial approach is more holistic than a purely trauma-focused psychology approach. Indeed, there has been some refinement in the recommendations for "best practice" in the years since my fieldwork. Neil Boothby, Alison Strang, and Michael Wessells (2006), for example, detail what they call social ecological approaches to children in war zones. The Inter-Agency Standing Committee (IASC) Guidelines on Mental Health and Psychosocial Support in Emergency Settings (2007) show that there has been movement away from a purely trauma-based approach to psychosocial work with former child soldiers. Psychiatrist Lynne Jones, discussing underlying principles to responding to the needs of children in crises, particularly those suffering during and in the immediate aftermath of conflicts and natural disasters, emphasizes "the need for attention to the child's perspective as a starting point and argue[s] for a deep consideration of culture, context and the specific meanings of events, as the framework both for assessment of the problem and response" (2008, 302). These changes to the recommended approach are heartening and have taken place since the time of my initial fieldwork.

During my fieldwork, what "psychosocial" meant in practice was unclear. On the ground it often ended up meaning tetherball, organized sports, art therapy, drama, or any kind of organized activity. At the time of my fieldwork, psychosocial programming was the least well theorized aspect of life in an ICC, and indeed this is the area where we find the biggest disconnect between what the West thinks is needed for child soldiers and what Sierra Leoneans think is needed. Michael Wessells agrees to a certain extent, concluding that "although Western-defined issues such as trauma often garner the most attention, child soldiers may also experience various issues that are culturally constructed and defy Western models" (2009, 588).

Traditional Healing

Another strand in Western programming is the demand for traditional "healing ceremonies." Much of this comes from "lessons learned" in Mozambique and elsewhere (see especially Honwana 1997, 2001, and 2006). There are many examples in the literature of refugees and other war-affected people who are better healed by the use of culturally appropriate treatments than by Western models such as PTSD. Anthropologists Edward Green and Alcinda Honwana in a World Bank report on the "Indigenous Healing of War Affected Children in Africa" conclude that "such disorders are in fact quite treatable by traditional healers, based on indigenous understandings of how war affects the mind and behavior of individuals, and on shared beliefs of how spiritual forces intervene in such processes" (Green and Honwana 1999, 3; see also Gibbs 1994; Henry 2000; Boyden 2000; Wessells 2006, 151–153). John Williamson of USAID is also a big supporter of traditional cleansing and healing ceremonies in the Sierra Leone context, mainly citing work from elsewhere (Williamson 2006, 196).

In Sierra Leone immediately after the war, NGOs were on the lookout for traditional healing ceremonies to support.[14] For example, when I met with representatives of the International Rescue Committee (IRC, an international NGO) they were very interested in finding out about such ceremonies so that they could be supported. However, when I asked Sierra Leoneans about such ceremonies, they almost universally denied that any such thing really existed. The most they would do with a returning child was to go before the village elders and inform them that a child was returning, more in the realm of political ceremony than spiritual ceremony. This is not to say that there were not postwar ceremonies, there were many. But they were more often to appease or commemorate the spirits of the dead, not to detraumatize or de-initiate.[15]

Although a proper orientation toward "local solutions" is key to the answer, why must those solutions always be cast in the rhetoric of "tradition" and "magic"? The argument for the existence of such ceremonies is often surprisingly mechanistic: if children were initiated magically, then there must be magic to de-initiate them (see, for example Krech 2003). This mechanistic notion does not come from any real sense of the culture; rather, it comes from lumping together all of Africa as one "traditional" mass. According to Alcinda Honwana (Honwana 1997; Green and Honwana 1999), such ceremonies arose naturally in the Mozambique context, but that is not the case in Sierra Leone. Based on fifteen months of ethnographic fieldwork in northern Sierra Leone, anthropologist Chris Coulter agrees that "there were no rituals or reconciliatory ceremonies that could 'wash' fighters from blood, death, and shame" (Coulter 2009, 248). However, outsiders have been looking for such a cultural silver bullet to the problem of reintegration from the beginning of their interventions.

Of course, Sierra Leoneans also strategically took advantage of this state of affairs. Not likely to look a gift goat in the mouth, communities accepted support for "traditional" ceremonies that sometimes involved the slaughter of an animal. I heard tell of a Kamajoh commander who, when asked if there were a "de-initiation" ceremony that could be performed, said, "Not really, but I can come up with something for the right price."[16] It remains to be seen how this strand of programming will play out in the years to come.

Schooling and Skills Training

Whereas "trauma" was a contentious term, one element of programming for child soldiers that Westerners and Sierra Leoneans easily agreed on was the need for schooling. It was not just programmers who pushed for education; former child combatants, parents, other community members, and NGO staff also interpreted combatants' reintegration needs as education or skills training (Women's Commission for Refugee Women and Children 2000). This can be explained partly by the common view that the lack of educational opportunities for youth was one of the main causes of the war (Richards 1996). Therefore, providing education for young people, especially ex-combatants, was seen as key to ensuring a peaceful Sierra Leone.

The formal education system in Sierra Leone was weak before the war. Low investment in education, especially outside Freetown, meant that only an estimated 37 percent of the school aged population attended school in 1985 (Government of Sierra Leone 1992). During the war, expenditure on education was minimal, averaging roughly 1 percent of the national GDP between 1998 and 2000 (UNDP 2003). In addition, hundreds of thousands of students and teachers were displaced, and large numbers of schools were destroyed and looted. As a result, at the end of the war, the Government of Sierra Leone estimated that 68 percent of the population between fifteen and twenty years old and some 500,000 young people between the ages of ten and fourteen had never attended formal school (Government of Sierra Leone 2000, 2003). These figures vary dramatically by region. The Government of Sierra Leone reported in 2000 that the Western Region, home of the capital city and the most accessible region of the country, had a dramatically higher enrollment (75 percent of school aged children enrolled) than the other regions. In the south (48 percent), the location of Pujehun, one of my field sites, and the east (35 percent) less than half of school-age children were enrolled, while only one in four was enrolled in the north (28 percent), the location of Masakane and Rogbom, two other field sites (Government of Sierra Leone 2000).

The UNICEF-sponsored Complementary Rapid Education Primary School (CREPS) and the Norwegian Refugee Council–sponsored Rapid Response Emergency Education (RREP) are programs designed specifically for students who had no access to formal schooling during the war. An example of the marketing for CREPS was found at the ICC in Kenema: a poster on the wall read, "*Complementary Rapid Education Program for Schools se big pikin no foh shame foh lan book*" (CREPS says even a big child should not be ashamed to learn). Initiated in 2000 and phased out in 2002, the Rapid Response Education Program (RREP) was a six-month program targeted at internally displaced and refugee returnee youth ages ten to fourteen who had no or limited access to formal education. The program was meant to support their ability to reenter primary school; it emphasized numeracy, literacy, trauma healing, peace education, human rights, health and physical education. In 2002, RREP was merged into the CREPS program. The CREPS program targets over-age (ten- to fourteen-year-old) out-of-school children and provides them with a six-year primary education that has been condensed into an intensive three-year program.[17]

The Sierra Leone government implemented free education for classes 1–3 in 1999-2000 and expanded free education to classes 4–6 in 2000-2001. Local communities have responded and enrollment has swelled in schools throughout the country (Women's Commission for Refugee Women and Children 2004, 62). A 2007 World Bank report on the state of education in Sierra Leone cites quickly growing enrollment figures for both boys and girls at the primary level (Wang 2007, 38). Although school fees have been eliminated, some Sierra Leoneans are still unable or unwilling to send their children to school because of numerous hidden fees, or because they need their children's labor .

UN agencies, the World Bank, and a variety of other donors, international NGOs, and private companies are active in rehabilitation of the education system in postwar Sierra Leone.[18] The many organizations and projects have greatly helped Sierra Leone in the past few years as well as raised hopes for the rebuilding not only of the education system but also of the rebuilding of the economy and society. Even with these achievements, the multiplicity of efforts and funding have raised concerns about coordination and fair distribution of resources, and the sustainability of inputs to education in Sierra Leone (Women's Commission for Refugee Women and Children 2004, 63).

Education in ICCs

Perhaps unsurprisingly, almost everyone I talked to thought that education was a good remedy for child soldiers, and it was a main programming element of ICCs. The form of that education was interpreted rather narrowly as either a return to formal schooling—often through a detour into RREP or CREPS first—or skills training for those who did not want to, or could not, go to school.[19] (The main programming element for adult ex-combatants was also skills training.)

The nature of education and skills training offered varied across ICCs. At many ICCs children attended makeshift schools operated within the

Figure 2.3. Children in a makeshift school at the Port Loko ICC.

confines of the ICC. Often, staff members at the ICCs were unemployed teachers. Unlike normal school, children in these makeshift schools did not wear uniforms. These schools often made use of the RREP or CREPS curricula, and sometimes they would use the RREP to determine the appropriate class level (for example, a student might be old enough to be in class 5, but since he had been in the bush for five years, it would turn out he should be in class 3.) This was the first time many of these children had ever attended anything even resembling school.

SKILLS TRAINING IN ICCS

Skills training in ICCs was divided by gender. Boys had carpentry, tailoring, auto mechanics, masonry, and other skills. For girls there was gara tie dyeing,[20] soap making, and hairdressing. Although most women in Sierra Leone make a living in farming and petty trading, those skills are already so ubiquitous as not to need training. Skills training took different forms in ICCs. Occasionally, workshops were set up on the grounds of the ICC (carpentry at Lakka, tailoring at Lungi). Sometimes skilled practitioners were brought in to demonstrate their skills (a basket maker at Port Loko). The idea of skills training was to provide children with skills with which they could make a living after reintegration.[21] Skills offered were determined in advance by the NGOs based on what was available or standard, not necessarily based on the best possibilities for post-reintegration livelihoods.[22] Sometimes, the training was seen as psychosocial, simply an activity that children could do together with each other and with community members.

Family Tracing and Reunification

One of the most difficult jobs of the child protection workers was family tracing and reunification (FTR). There were many challenges. First, approximately three-quarters of Sierra Leoneans experienced displacement at least once during the war (Abdalla, Hussein, and Shepler 2002). This made it difficult to locate people even when a child could remember the details of his or her family life before war. Sometimes, children would tell differing stories about their origins either out of fear of reprisals or in strategizing about moving to an ICC with better benefits. Sometimes, when NGO workers were able to track down relatives of former combatants, the family refused to accept the child back, either due to poverty or to unwillingness to forgive the child for atrocities committed during wartime. Often, a child would decide he or she was better off in a relatively well funded NGO program than back in a village with few educational or vocational opportunities. Many times, a child came from an area that was under rebel control and therefore inaccessible to the NGO workers.

On the other hand, the job was made easier by certain aspects of Sierra Leone culture already covered in chapter 1. The strength of the extended family system meant that even if a child's parents could not be located, the child could often be reunified with more distant relatives such as aunts, uncles, or grandparents. Also, the value of a child was such that families might be willing to take on a child not their own (especially if they were also promised some forms of help from the child protection NGO). This too falls in line with the fosterage system discussed in the preceding chapter. As I discussed in the introduction, some NGOs were even reunifying children with extended families in internally displaced persons camps like the one in Jerihun.

FTR did not always happen solely as a result of NGO workers moving out into the countryside to find families. Occasionally, families missing children would show up at ICCs looking for their kin. There were also new forms of FTR being tested near the end of my fieldwork. For example, a radio tracing program had begun wherein descriptions of children in ICCs were read over the radio and people having connection to those children were asked to contact the appropriate NGO. There was also photo FTR, in which NGO workers would go around camps and show the pictures of children in their ICCs and asking if anyone knew the children or their families. There was also increased activity around international FTR in refugee camps, mostly carried out by the International Committee of the Red Cross. They would try to help separated children who were in camps in Guinea, Liberia, and even the Ivory Coast to find their families back in Sierra Leone.

There were many safeguards built into the system. NGO workers were very careful about making sure they had the right family. They would interview the family about the date the child was taken, the age, and any distinguishing marks before making a positive ID. They would talk not just with the family, but also with important people in the village about the return of a child. In some cases, a family might be happy to receive a child back, but the community might be threatened by that child. The NGO workers often had to do a lot of "sensitization" of community members about forgiving the child. This is where an NGO worker would tell community and family members, "He was only a child. It was not his wish to fight. We must accept him back. *Bad bush no de for trowe bad pikin.*"[23] They would similarly counsel the children about how to behave on their return, sometimes acting out skits of what reunification might look like.

When the system worked as it was supposed to, it was a beautiful sight. One of the most moving events of my time in Sierra Leone was a reunification I was lucky enough to witness in Port Loko. I was visiting the ICC for a week, just hanging around talking to people, when the head of psychosocial programming called me over. He asked if I would like to go along on a reunification. He had a father with him who had come by the ICC on the chance that his daughter might be among the children at the center. It turned out the girl was staying with a family on the outskirts of town in a fosterage arrangement. On the way to the house, in a big white Land Rover, the father was very excited. He told me he had not seen his daughter in two years. He had a horrible story to tell about how they were separated. He was on the farm one day when rebels attacked his house. He got close enough to the house to overhear them saying that they had killed everyone there except one little girl. He heard them debating whether to kill her too, or to take her along with them. He stayed hidden throughout the discussion in fear for his own life. They eventually decided to take the girl with them. The father was pleased that at least her life had been spared. He had been looking for her ever since. When we arrived at the foster house, the family was expecting us. There were big grins all around as the man came down from the car and the girl recognized her father immediately. She cried out "Papa!" and ran to his side. They stared at each other in disbelief, not speaking. They stood in shock with tears running down their faces and down the faces of the foster family with whom the girl had lived for about a year. Everyone gathered on the veranda, the girl and her father inseparable. The father thanked the family for taking such good care of his daughter, and they responded that they were just so happy to see them reunited. The girl led her father by the hand to show him the room in the house where she had been staying. All of us—observers, NGO workers, and members of the foster family—were overwhelmed by the emotion of the event. After a time, the NGO workers and I went back to the ICC with the father and left the girl behind to say her goodbyes. The psychosocial programmer told me that in his experience, sometimes reunification could be as traumatic as separation. There were so many questions, and often people were afraid to hear the answers.

Sometimes, when FTR did not work, children would take their situations into their own hands. At the Lakka ICC I met a girl named Fatu who had been with the RUF for some time and had been the "wife" of a high-ranking RUF commander. Many of the boys who came into the ICC recognized her and respected her. When I visited the Port Loko ICC, I was surprised and happy to see her there, sitting in the main compound, working on a craft project with some younger girls. She told me that she was sure her family was in Makeni but that none of the NGOs were doing anything to try to find them for her. The reason was that Makeni was still under RUF control at that time, so NGOs were not operating there. She was not afraid of the RUF, and she wanted to know if her family was in Makeni or not. She had managed to get transferred to the Port Loko ICC, the closest to Makeni, and she was gathering funds to make the trip there on her own. (She told me this but asked me not to tell the ICC staff.) She wanted to find her family in Makeni and had to work around the system to do so.

The Ideal Trajectory

Rehabilitation and reintegration programs in Sierra Leone assume a standard trajectory in the life course of a child soldier: normal village life, abduction, "in the bush" (fighting), demobilization, rehabilitation, family tracing and reunification, reintegration (and therefore back where he started). He (for the assumed child soldier is usually male) was living a regular life in a regular village when he was abducted and forced to do unspeakable deeds. After some time, due to the pressure of child protection agencies, he is turned over for rehabilitation and reintegration.

The child protection agencies use a model that works like this: While children were captive, they were passive victims. Now that they are demobilized, they are choosing what they want to be, getting training, and doing something forward looking. There is an assumption for all child ex-combatants that they go from one type of space to another:

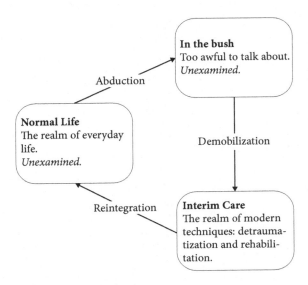

Figure 2.4. The ideal reintegration trajectory.

there they had no agency and were not learning anything, *here* they have agency and are doing something better for their future. The assumption is of a clear break, yet, in many ways in practice, the break is not that clear. Children were dealing with some of the same daily struggles in interim care centers as they were elsewhere. There was still structural and symbolic violence in their everyday lives.

Conclusion

My objective in this chapter was to describe the interim care system for former child soldiers along with some of its ideological underpinnings. The goal was to describe it not as a perfectly functioning system but as one that built on previous systems in other countries, being designed on the fly by child protection workers of good will. In the next chapter I describe some of the ways people, Sierra Leonean and expatriate, child soldiers, villagers, and city dwellers, maneuver through and within that system, in the process *making* the "child soldier."

3

Learning "Child Soldier" across Contexts

The "Auto Biography" below was given to me by a worker at Children Associated with War (CAW), a Sierra Leonean child protection NGO based in Freetown and active in, among other places, my field site Pujehun. He wanted to show me some of the good work his NGO was doing with ex-combatants. (I have changed and deleted some identifying details.)

Auto Biography

My name is Aiah Mbayo. I was born in 1977, in Koidu Town, Kono District in the Eastern Province of Sierra Leone. I come from a family of six with three brothers. My father Sahr Mbayo was both a farmer and a miner and my mother Sia was an ordinary house wife. I started attending school at the age of 7 years.

In September 1991 whilst we were together in the farm with our parents, a group of rebels of the Revolutionary United Front (RUF) arrived and opened fire on us, killing my father, three brothers and an uncle. I was captured and taken to the bush, and trained to fight, and eventually made a commander of the Small Boys Unit (SBU) of the RUF. I stayed with the RUF rebels for nearly 6 months around Bunumbu Teachers College axis, until when we were attacked by the Government troops with heavy firing killing 50 men out of 200 and captured 32 militia boys including my self with minor wounds. We were taken to Government controlled Garrison, town of Daru Military Hospital. I stayed in the hospital for one week. Whilst in the hospital I thought of the brutal killing of my father, brothers and uncle, so I developed a sense to revenge on the rebels.

I joined the Sierra Leone Army as a child Soldier. Whilst staying with them at Daru Barracks, the RUF rebels attacked the Barracks. The battle lasted for four days but finally the rebels were repelled. Daru was my first place off (*sic*) serious battle with the RUF and this marks the beginning of my military life. I fought in so many towns/villages like—Baiima, Jojoima, Mobai, Kuiva, Dodo, Nyandehun, Pendembu and Kailahun.

In the battle field, they called me "KILLER." I stayed with the fighting forces for nearly two years using various weapons like AK 47, G3 Anti-Air Craft Gun (A/A) etc.

In June 1993, when CAW was established, I was amongst the first set of children (360 boys and 10 girls) enrolled in the Programme and stayed in one of the centres called Benin Home, at Wellington, Freetown to under go counseling and psychosocial therapy. After spending three months in the Home, I was taken away by a senior military officer to go and fight again. I spent two months at the battle front before coming again to the home in December 1993 when I was placed with a foster parent in Freetown.

I gained admission to the Prince of Wales Secondary School, May Park, Kingtom and was placed in Junior Secondary School Class / Grade 1 (one) in January 1994 until when I finally sat to the Basic Education Certificate Examination (BECE) and the Examination West African Senior School Certificate Examination (WASSCE) respectively and passed. I have applied for the Association of Chartered Certified Accountant (ACCA) for course which I am presently looking out for sponsorship.

Though Aiah's story is certainly compelling, I could not help thinking that what had really been achieved in this document was a kind of marketing of Aiah's story. Aiah had learned how to identify himself as a "child soldier" for a Western audience, with the hope that he might get some financial support for his education as a result. He even had a Yahoo e-mail address to facilitate the process. This is not to say that this was all put on, or even that Aiah did not deserve some help with his schooling. Rather, I am pointing to the ways that Aiah's strategic self-presentation was influenced by Western models and NGO practices.

In Sierra Leone, "child soldier" is made at the intersection of local and global models of childhood, in social practice, between partially determining structures (in this case, current and historical Sierra Leonean practices regarding childhood and youth and the Western model of childhood) and personal agency (in this case, children and adults using the Western model of childhood for their own purposes). "Child soldier" is made in and around institutions in multiple and sometimes contradictory ways. The ideological underpinnings of these institutions is a Western, individualistic framework, yet the actual effects are to be found in Sierra Leoneans interacting with (making and remaking) the institutions; that is, the effects are in social practice. In general, the steps involved in making "child soldier" are as follows:

1. An external distinction is imposed (imperfectly).
2. People strategize regarding that distinction—in part by deploying history and local meaning, also taking advantage of the confusion in administration of the distinction.
3. The distinction comes to have local meaning and the struggle transforms the external distinction.

As is probably already clear from my description of the Bo ICC, the system-as-designed broke down in many places. One source of these breakdowns was the philosophical and practical problems with the design of the ICC system from the beginning, and the fact that the system was being put into place on the fly, in an always-unpredictable conflict-affected setting. But it also broke down due to the maneuverings of individuals who participated in it in unanticipated and unintended ways that both helped and hindered their "reintegration." In fact, the trajectory for child soldiers is *not* unilinear and there are *not* clear breaks between the various assumed stages.

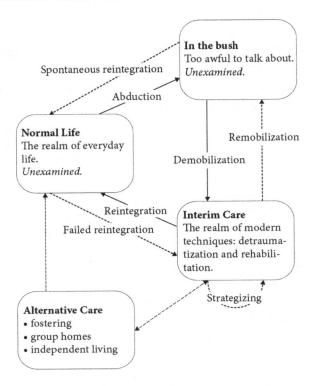

Figure 3.1. The modified reintegration trajectory.

Figure 3.1 starts with the ideal trajectory presented earlier but adds in some of the other possible pathways. For example, one may leave the bush and go home without participating in the ICC system. That path is represented by the dotted line titled "spontaneous reintegration." One may move from an ICC back to an old commander or even to another fighting force. That path is represented by the dotted line titled "remobilization." Moving around from one interim care center to another in search of the best benefits is the curved dotted line titled "strategizing." Children may move back and forth among ICCs, "home," and any number of possible alternative care settings. Generally, rather than one predetermined circuit from normal life, to the bush, through an ICC and back to normal life, it is possible to move all over this map from any state to any other. Throughout this process "child soldier" itself becomes a useful identity.

The physical boundaries of ICCs were less sharp than I expected. Children had to go in and out to go to school and to get water. People wandered in and out constantly: orange sellers, even knife sellers. As a foreigner, after an initial questioning by the people in charge, it was easy for me to hang out informally all day with the kids. NGOs had far less control of the physical movement of children and ex-combatants that I had originally imagined. Children would transfer themselves to the program perceived to offer the best treatment (such as food or entertainment). Because children compared the centers and strategized about where to get the best package, it is important to understand the centers in relation, rather than taking any one in isolation. Since children moved around the country through a system of institutions, I was almost guaranteed to meet someone who knew me when I visited any ICC for the first time. I was pleased, for example, to be greeted on my arrival at the Bo Center by several children who recognized me from their earlier stay at Lakka.

The Lakka Interim Care Center

The ICC at Lakka is the one I visited first and the one I know best, since I spent many months there and continued to visit it throughout my fieldwork. The Lakka ICC served the Western Area, including Freetown. It was located right on the beach in a town along the peninsula outside Freetown in what had been a mid-level tourist hotel before the war. There was sand everywhere, and palm trees bending low over the beach. The compound was made up of a large two-story building that had been the original hotel ("phase one") and the more recent addition of a series of smaller buildings called bungalows surrounding a central communal area ("phase two"). There were additional bungalows farther away ("phase three") that were not part of the ICC that housed internally displaced persons. There were always construction projects going on to shore up the old phase one building and to remake the rooms to suit an evolving organizational design. Most of the time I was there, the administration and a room for the few girls were located in the main building and each bungalow was assigned a group of boys and a caretaker, making up little social units.

Lakka was run by an Italian Catholic priest with many years experience in Sierra Leone. He and his people (Sierra Leoneans for the most part) had turned themselves into a local NGO named Family Homes Movement so that they could run the ICC. They had previously run a street kids program in the east end of Freetown. I had many long discussions with him about his theories of what was needed in Sierra Leone. He saw his organization as rescuing children from a life of deprivation. As a result, and contrary to the Ministry guidelines, some of his charges had been at the ICC for up to three years. He was in no rush to reunify children where they might face hardship or rejection from their communities. Lakka appealed to Westerners' ideas of innocent childhood, situated as it was right on the beach. An American working for the UN Human Rights section remarked to me while visiting the ICC that seeing the children play on the beach at Lakka always made her feel that "at last they can have a normal childhood." But playing on the beach all day is *not* a normal childhood for Sierra Leonean children. In fact, it was somehow galling to Sierra Leoneans to see children, rebel children at that, treated so well.

Lakka was more than a home for former child soldiers and staff and was always abuzz with activity. There seemed to be a large population of hangers-on, particularly women and children, who had some connection to the Catholic priest who ran the place. There were various projects in development: a fledgling piggery, a garden, a carpentry shop, a fishing boat, and a traditional loom. Two large water pumps in the central courtyard of phase two were a gathering place for people to do laundry, gather water for bathing, and gossip. Cooking was communal and took place in a large open kitchen. When it was time to eat, a representative from each bungalow would go and take a big plate of food for the whole group to share. Outside the compound was a large paved area (perhaps originally a parking lot) that had been turned into a perpetual football pitch. Also popular was what I came to think of as the official game of ICCs: tetherball. Most Sierra Leoneans do not know how to swim, but the boys would sometimes make brave ventures into the surf and run screaming in fear and glee when the waves crashed near them.

I was at Lakka about three weeks after my visit to the Bo ICC I introduced in the previous chapter. Who should I meet relaxing by the beach but my pal, Political Peter. He seemed genuinely happy to see me again. I bought some fruit from a passing vendor and he and I shared it over his story. It seems the other boys he had run away with had decided to try life on the streets of Freetown, but he had heard good things about the Lakka ICC. He showed up on their doorstep and had been taken in. He was enjoying life at Lakka and thought he might stick around for a while. He especially praised the quality and frequency of food, pointing out, "We get rice every day, and we get tea in the morning!" All he needed in order to stay was to fabricate a fictional family in Freetown he wanted to be reunified with. Peter is an example of someone who was—at least in his own mind—making the most of the system.[1]

The centers were different, and, interestingly, the people working in them did not know this. Unlike the staff, who generally stayed put in one workplace, I followed the children around and found that children knew more than the staff did about the differences between centers and, more importantly, between the different localized meanings of "child soldier."

Psychosocial Programming at Lakka

The children in every ICC spent a lot of time talking about what was owed them, and comparing experiences, and even pumping me for information about other ICCs and benefits and sometimes asking me outright for a "morale booster" (money), the same way they had demanded money from "civilians" during the conflict. The most experienced RUF fighters had a sense of entitlement and would threaten staff and members of surrounding communities. Despite their frightening demeanor, many of the children at the ICCs were, in fact, suffering. One of the caretakers at Lakka told me that several of the boys in his bungalow woke up scared almost every night, asking nervously about Kamajohs or ECOMOG in the area. He explained the bad dreams and the continuing drug use among some of his charges by the fact that they were "traumatized," using the novel English word amidst his Krio. Indeed, part of what ICCs did was teach children that they were "traumatized," as one example of a drama performance organized at the Lakka ICC will demonstrate.

One day at the Lakka ICC, I watched a man hired by COOPI, an Italian NGO affiliated with the ICC, to organize the kids into a dramatic performance. He was a drama teacher from Freetown moonlighting at COOPI.[2] He collected about twenty boys who were interested in his project. He had them sing in a mixture of Krio and English (which I call Kringlish) to the tune of a well-known Krio song:

> Kaboh [Welcome] COOPI.
> We say welcome.
> Welcome to COOPI na [at] Lakka.
> COOPI dohn bring [has brought] one treatment
> For mehn we trauma [To cure our trauma].
> We foh [should] talk to COOPI so that
> COOPI go help we [will help us].

Then they broke into a play, scripted on the fly by the drama teacher.

Kringlish

BOY 1: Mi na pikin. Gi mi chance foh gro.
BOY 2: Mi na pikin. A get foh live tumara.
BOY 3: Wi thri na pikin, wi get foh de tumara.
1ST GROUP: Usai una komot?
2ND GROUP: Wi na Sierra Leoneans.
1ST GROUP: Wisehf na Sierra Leoneans.
ALL: Wi all get foh mek Sierra Leone behteh. Wi na pikin den. No gi wi gun, no gi wi kohtlas. Gi wi buk en pen.

English

BOY 1: I am a child. Give me a chance to grow.
BOY 2: I am a child. I will be alive in the future.
BOY 3: We three are children. We will be here in the future.
1ST GROUP: Where are you all from?
2ND GROUP: We are Sierra Leoneans.
1ST GROUP: We too are Sierra Leoneans.
ALL: We all have to make Sierra Leone better. We are children. Don't give us guns. Don't give us machetes. Give us books and pens.

The theme of the song is: we are traumatized, but the NGO will help cure us. The skit aims to: foster national identity, show that children are the future of the nation, and emphasize the right of children to education.

The children were meant to perform their song and skit at the beginning of a program featuring a professional dance troop and some more skits performed by professional actors hired by COOPI. After practicing all afternoon, the children and the drama teacher took a break while they waited for a dance troop to show up. I took the opportunity to talk more with the drama teacher about the project over a soft drink. I told him I really enjoyed the way he talked to the kids and that I liked the program because the young kids got to learn a song and take part in an activity they could feel proud of. He told me he enjoyed the work and it was a good way to make ends meet since teacher salaries were notoriously unreliable. He said that his colleagues wondered whether he was afraid to work with rebels, but he cast his participation in the program in almost patriotic terms, saying it was important for the future of the nation.

The dancers finally arrived, along with a few Sierra Leonean COOPI staff members, including the director of psychosocial programming. The program was clearly based on similar programs they had done at displaced persons camps around Freetown, and some of the skits were aimed at an adult audience (for example, skits against wife beating and alcoholism). It seemed hard to hold the children's attention during the performance, and some of the boys were openly derisive. When I commented that they did not seem to be paying attention, the COOPI staff member said the message would sink in with time and they would remember the program later. The dancers were exciting, and the people of the ICC—adults and children—clearly enjoyed them. In the skits, it was interesting how the NGO was anthropomorphized and lauded: "*Talk to COOPI. COOPI go solve wi problems, wi wae traumatize*" (Talk to COOPI. COOPI will solve our problems, we who are traumatized), echoing the theme of the children's song that we are traumatized, but that COOPI will be able to solve our problems.

It is unclear what kind of impact these kind of psychosocial activities had; I saw similar activities at all of the ICCs I visited.[3] The children clearly enjoyed the diversion from their daily routine in the centers, but were they "detraumatized"? It seems to me they were learning lessons about the *language* of trauma and what their relationship to NGOs should be.

Discourses of Abdicated Responsibility

In Sierra Leone, the newly imported idea that anyone under eighteen years is a child and therefore not to be held accountable now allows whole groups of young people to be forgiven by their communities in a new way. This obviously helps the young people who are struggling to reintegrate; it also helps the communities into which they are moving. The primary way in which the discursive object "child soldier" is used in Sierra Leone on a daily basis is through discourses of abdicated responsibility: people say things like "they were all on drugs," "they were all abducted," "they were just kids and didn't know what they were doing." In addition to stories of wartime suffering, the children use these discourses to negotiate their acceptance and readmission to society.[4] Society uses these discourses to smooth their readmission too, as well as to explain to themselves how such a horrible thing could have happened. These claims of innocence ease children's reintegration into communities and make it easier for community members to live with former fighters in their midst. Adult combatants use some of the same strategies, of course, but there is something quite specific to the case of children.

The child soldiers I met in my work are navigating a very tricky social landscape as they move through various intersecting contexts. Among their friends and fellow soldiers, they try to maintain the status that being part of the fighting gives them. They wear combat clothes and sunglasses, and brag about firing rocket-propelled grenade launchers. With NGOs they adopt the persona of the traumatized innocent, usually requesting help in furthering their education. With community members and in school they try to act like normal kids, never mentioning the past. Thus their "reintegration" is achieved in social practice across a variety of contexts using a variety of strategically adopted identities.

Child ex-combatants exercise agency, paradoxically, through their claims of wartime *nonagency*. Youth in this postwar context are strategic and skillful users of these different discourses as they move through different contexts. For example, I have often seen ex-child combatants on their own accord manipulate their image for the media; that is, they run and put on their rebel sunglasses and bandanna (or Kamajoh traditional garb) when a photographer is present. Child rights discourse and practice in some ways ease the reintegration of child ex-combatants by buttressing these "discourses of abdicated responsibility" in children's narrations of their war experiences, thereby facilitating community acceptance.

Skills Training at Lakka

The skills training program I know the best is the gara tie-dyeing class at the Lakka ICC, because I was a dedicated participant. As a woman, I could not really learn any of the "male skills," but my gender and my ignorance made it totally acceptable for me to join the gara class. The class was organized for girls in the ICC as well as women working at the ICC and women from the surrounding community. The ICC brought in two displaced gara experts from Makeni to teach the class.

I enjoyed learning how to do gara dyeing, and it gave me new appreciation for styles I saw in the market and street. Most importantly for me, participating in the tie-dyeing class gave me a wonderful vantage point from which to participate in the goings-on at the ICC. The girls in my class saw me struggling to learn just as they were. I became, for the ICC community, a member of the gara class, not another NGO worker (which is how most white people were seen). I knew I was somewhat successful at blending in when an Italian woman related to one of the Italian priests who ran the place was visiting, looking at the finished products of our class packaged for sale to the outside world as the work of former child combatants. The priest introduced me to his mother as "Susan, a member of our gara class." Our teacher, Auntie K, sent me and one other girl to her bungalow to retrieve some more samples of our output. Clearly, I was not an ex-combatant, but it pleased me to be treated like just another member of the gara class, and I believe it pleased my classmates. I find it ironic that a few Italians somewhere own some inexpertly tied gara they believe to have been produced by a traumatized Sierra Leonean girl, in fact produced by a clumsy American anthropologist.

Figure 3.2. My gara tie-dyeing group.

Despite a rise in the popularity of "culture" clothes"[5] in postwar Sierra Leone, there is no way that Sierra Leone can absorb all the gara dyers trained in such programs all over the country. The skills training provided in the ICCs was not substantial enough for anyone to become a skilled practitioner, and when children left the programs they did not have the tools needed to make a living at carpentry, tailoring, auto mechanics, or whatever the skill they had been taught. In my opinion, there tended to be an overreliance on skills training as part of the reha-bilitation package. In the years since my fieldwork, policy makers have acknowledged this problem and have generally treated it as a problem to be solved by market assessment (Beauvy-Sany et al. 2009).

Reflecting back on the practice of apprenticeship discussed in chap-ter 1, it should be clear that skills alone are not enough to gain a liveli-hood even with the right market assessment. In Sierra Leone, the social relations between a master and an apprentice are vital to ensuring a successful outcome. Some of the least effective programs I saw were skills-training programs set up like schools, growing out of the Western model of skills separated from social relations. Children coming out of such programs could not use the skills they had learned, and I suspect they knew that going in. Rather, they thought they might be able to get something like a formal education that they might parlay into further formal education, and this idea had resonance because of the particularly Sierra Leonean valorization of formal education.

Although skills-training was often not that useful, it sometimes led to apprenticeships that were useful. I saw some effective apprenticeship fosterage relations at the ICC in Lungi, wherein children from ICCs were fostered to masters of a trade in the traditional way. For this apprenticeship model of skills training, the ICC staff found people in the community who could train apprentices in carpentry, auto mechanics, masonry, and other fields. Then they provided some tools and food to "pay kola" (pay a small symbolic fee) for the training. Children lived and worked with the master in his household. Not only did they learn skills, they also got the "blessings" of a master, and integrated into a normal Sierra Leonean household.

Education at Lakka

Lakka was unusual among ICCs because children who wanted to attend school were sent to a school in the nearby community, Sengbe Pieh,[6] instead of attending a makeshift school inside the ICC. With respect to their eventual reintegration, this situation had advantages and disadvantages. Although they were mixing with the local population, they were much less likely to want to leave Lakka to be reunified in villages with no educational facilities.

The vice principal at Sengbe Pieh was born in the village I had lived in for two years as a Peace Corps Volunteer, and he was always happy to help me with my research. He and I had many long discussions about the school. He characterized the child ex-combatants in the school as disrespectful of elders, disruptive, and lacking concentration and focus. However, he was glad they were mixed together with the regular students, pointing out that "we have all suffered from the war." He complained that the "normal" children were influenced by the ex-combatant boys. He gave the example of a boy being *fit yai* (disrespectful) to a teacher in front of the whole school, saying the other students would surely look up to him and say he was a man. He explained, "The problem with our children is that they copy everything that is bad rather than what is good." I asked whether the "normal" children had any influence over the others. He said, "Of course. When a minority wants to be part of the majority, be part of their society, they just have to copy what they do." He admired the unity of the ex-combatants, saying, "You will really know they are united if someone wrongs one of them. They all stand together to defend him."

The home economics teacher at the school was a friend of mine.[7] She was originally from Kono in the east; when rebels attacked her compound and killed her father, she and her mother fled to Guinea and lived in the bush for six weeks. She ate only bananas during that time and had to be admitted to the hospital afterward. I asked her about the ex-combatant students at the school. She admitted they were quite bright in general and that they were the top five students in form 1 (equivalent to the first year of junior high school, or seventh grade). However, they did not want to take discipline. She told me about one boy who grabbed the cane from a teacher as he was about to be beaten. She told him, "You were the ones who wanted to go to school. Didn't you know that they beat in school?" I asked if the ex-combatant kids mix with the regular kids. She said yes, but the problems come when they have a girlfriend in common. Then the rebel boys all stick together.

The principal, Reverend Bendu, said most of the ex-combatant children had no problem adapting to the society of the school, though one or two caused problems. He had to tell one boy to find another school. Of course, he admitted, other kids had discipline problems as well. He stressed the importance of keeping all the children together and not enforcing any separation ("as they do in white man's country"). He told a story about one boy who was making fun of one of the teachers who is crippled from polio. He called the boy into his office and said, "When you first came to us, did anyone make fun of you? Did anyone call you a rebel? Don't you think teachers have feelings too?" After that the boy never did it again and in fact wrote a letter of apology. Reverend Bendu told me that when the school first admitted ex-combatant children, he made a speech at assembly urging everyone to treat them the same, explaining that it had not been their choice to fight. He argued that his teachers were doing the important work of helping bring these children back to normal, and that they should get some extra compensation for that. He concluded that the few children who had some "training" (that is, home instruction in proper behavior) before they went to the bush were easier to bring back to society, and the ones who had known no training beforehand were the most problematic.

Convergence of Desire around Schooling

Why were schooling and skills training so popular, despite the general lack of economic opportunities for graduates of schools or even of skills training centers? The fact that young combatants often preferred formal schooling makes sense within the context of the history of education in Sierra Leone. Robert Krech (2003, 140) cites conversations with NGO staffers who told how former RUF child soldiers demobilizing in the remote and devastated communities of Kono district said they wanted to become computer engineers or pilots, and I had many similar discussions with ex-combatants myself. Michael Jackson, a longtime ethnographer of Sierra Leone, calls the ex-combatants' unrealistic career expectations "the mystique of literacy." He recounts stories from his fieldwork at Kabala Secondary School in the north in the late 1960s, pointing to "the poignantly impossible gulf between [students'] dreams and their reality. Though most were the children of farmers, they showed their disdain for farming in the zeal with which they laundered their uniforms, washed their bodies, [and] manicured their fingernails" (Jackson 2004, 148). Formal (or academic) education has never had to make sense within the political-economic context in order to be the marker of an elite "educated" identity. Hence we see the apparently impossible aspirations of an illiterate boy to become a doctor or lawyer through the rehabilitation program. The educational system in West Africa has always seemed mysterious and unequal to rural Sierra Leoneans, but a path to a hugely desirable future.

Sierra Leoneans overvalorize education and assume that once a person is educated he or she will no longer have to work, at least do physical work on the farm. According to non-Sierra Leonean West Africans I met, most of whom were with ECOMOG or the UN in some capacity, this is the Sierra Leonean malady. Their history of educational attainment has left them prizing only education at the expense of other values. There is in all of this an implicit devaluing of agricultural labor, the labor in which most most Sierra Leoneans are engaged. In chapter 1 I explained this bias as growing out of the colonial history of education in Sierra Leone.[8] This view has only been exacerbated by the example of educated government officials in the postcolonial period.

Look at the irony of the situation: the lack of opportunities for young people and the inherent structural violence of the educational system lead to "a crisis of youth" that leads to war. The proposed solution is to continue with the flawed ideologies of Western schooling and heal the young combatants with the almost magical application of education. Education has become the default solution for any problem of youth in the liberal Western framework. So the centrality of education to the ICC system emerges from the confluence of Sierra Leonean *and* Western misconceptions about the promise of education.

The Politics of the School Form

The various activities in ICCs reflect an underlying universalistic, individualistic ideology in their design. To us in the West, the assumptions behind childhood, schooling, and psychotherapy are hegemonic, so it requires a shift in thinking to see that these are ideologies and not, as the hegemonic would have it, free of political content. Schools—and school-like organizations like ICCs—are political institutions.

As anthropologist Jean Lave (2011) argues in the case of tailors' apprenticeship in Liberia, what is achieved in ICCs is not merely a transfer of knowledge. The participants are learning to inhabit new identities; they are learning to be former child soldiers across contexts in historically and geographically situated social practice. Essentially, this is an argument against viewing learning as merely knowledge transfer. The activities in ICCs are organized around the assumption that lessons will transfer to the post-ICC world. As in schools, everything that happens in ICCs gets meaning from what happens afterward, yet the people working in the centers do not know what happens later. Just as "word problems" posed in school settings only make sense in school settings, the problems posed in ICCs (for example the "reintegration plans" activity at the Bo ICC that asked children to recite what they would do when they were asked to participate in communal labor) only make sense in the context of the ICC.

Relations between ICCs and Surrounding Communities

The identity "child soldier" is made in relation to a set of formal institutions designed for their remaking, but relations with wider societal institutions are also important. The social relations underlying "child soldier" extend beyond the ICC network. In particular, we must look at the relationships between ICCs and surrounding communities.

The school attended by the Lakka ICC children is an excellent example of what I am talking about. Sengbe Pieh school is located in a small fishing village very close to Lakka and the ICC. The ICC was originally placed in the area without much consultation with the surrounding communities, with resulting tensions. I write more about the tensions at this school in Shepler (2005). Some people were afraid of having "rebel children" in such close proximity. They were annoyed that the "rebel children" were provided with international aid that supplied them with food and school fees when many of the community members were struggling to get by without such help. The local school was a main issue of contention. During the 1980s, with money from its lucrative communalized sale of beach sand to the construction industry, community members in Hamilton built the local school. In addition to serving the local people, the school was also attended by about one hundred child ex-combatants who lived at the nearby ICC. These students were mostly boys, ranging in age from ten to twenty, from class 1 (the first grade of primary school) to form 3 (the last grade of what is known as junior secondary school in Sierra Leone). Based on interviews with individuals and on several PTA meetings I attended, I know that the community had mixed feelings about the boys' presence in the school. They were often afraid of the ex-combatants. They also felt that it was not right that former soldiers who had inflicted so much suffering on so many innocent people should benefit from the school that the community built. On the other hand, this new population of child ex-combatants came with certain benefits, like the support of their sponsoring NGO and of UNICEF. In particular, the NGO was paying school fees for all the students it was enrolling. UNICEF had recently helped the school build a wall and get a water pump working, solely because the school was enrolling ex-combatants. Moreover, the principal, vice principal, and a number of teachers were employed at the school only because they had been displaced from their own schools by the civil war. Although the community tried to portray itself as the owner of the school, local financial support for the school over the years had been spotty at best. Simply put, it is likely the school could not have been operating without the financial support the former child soldiers brought in.

In 2000, these tensions escalated into violence. Some boys from the center got into an argument with an auto mechanic in town and broke an automobile windscreen. Some members of the community decided they had had enough of this sort of disrespectful behavior and decided to take matters into their own hands. After several days of tension, there were injuries and property damage both at the ICC and in the surrounding community. The word in nearby Freetown was that the rebel boys *don baranta* (had gone wild) and that several people were dead. After hearing the rumors, I went to Lakka as soon as I could to investigate. In fact, no one died as a result of the tension, more damage had been done by young men in the surrounding community, and most residents of the ICC were holed up, concerned about the possibility of further attacks.

Following that event, UNICEF and other aid agencies decided that there was a need for more "sensitization" with the surrounding communities in order to safeguard the work of the interim care center.[9] Some supplies donated to the neighboring communities by UNICEF—cooking pots and the like—had helped to mollify some of the community members, but many continued to argue against accepting the former child soldiers in their midst. "How can we be expected to help these children when we cannot even help ourselves?" they asked.

The relationship between ICCs and surrounding communities is not always so fraught with tension. For example, Caritas Makeni ran a very successful fosterage program at the Lungi ICC in which it fostered children from the ICC into the surrounding neighborhood, sometimes as apprentices. It took some convincing at first, but eventually some members of the community decided to take on former child soldiers as their own children. One woman put her reasons for fostering particularly eloquently. She said, "These children are the rebels' best weapon. If we take them away and retrain them, we have disarmed the rebels." Hers was a particularly poignant case. The rebels had killed members of her family and burned her house down, but still she decided to foster two boys. At first, like most of the community, she was against the idea of bringing them to her town. But eventually the sensitization program worked on her and she was able to convince her husband that they should foster a child.

Conclusion

This chapter has shown the ways people—Sierra Leonean and expatriate, child soldiers, villagers, and city dwellers—maneuver through and within the child protection system, in the process *making* the "child soldier." "Child soldier" is produced differently in different locations. It is not a top-down definition; rather, it is partly made though young people's own strategizing about location and about self-representation. Politically and materially, the identity "child soldier" carries with it a range of meanings and implications and serves as a site, both discursively and in the lives of the children themselves, for both the vernacularization of child rights and the reform of Sierra Leonean culture.

4

Informal Reintegrators, Communities, and NGOs

The very term NGO acquired a bad name during the UN
intervention, when the number of NGOs mushroomed
uncontrollably. There are still a good number of organizations that are acutely aware of the shifting agendas of foreign
donors and who rapidly adjust their own priorities accordingly. When 'women's issues' is the rule of the day, the foreign visitor will find a host of very articulate local organizations ready to take up work in that field. . . . As long as casual
visitors are ready to naively distribute funds without setting
up systems of accountability, these types of organizations
will continue to exist (Richards, Bah, and Vincent 2004, 27).

Rogbom is a small Temne village on the road to Freetown in Koya chiefdom. Anyone entering or leaving Freetown must travel along the main
highway near Rogbom. Freetown was strongly defended by the army
and by international forces. Therefore, the people of Rogbom were kept
out of the war except during times when Freetown was attacked. They
emerged relatively unscathed from the junta and the intervention in
1997. The war really affected them in 1998 and 1999 in the buildup and
the aftermath of the January 6, 1999, invasion of Freetown. In the last
months of 1998, the RUF was in control of about 80 percent of the country and had a new base in the capital of the Northern Province, Makeni.
RUF forces moved closer and closer to Freetown, taking over villages as
they progressed. Freetown people think of "January 6th" as a few weeks
of violent occupation. Rogbom endured many months of violent occupation, first by a movement building strength through abduction and
theft, and then by a disappointed and frustrated movement in retreat,
wreaking vengeance as they went.

When the rebels came through before January 6, everyone ran away to the bush, and stayed there for three months. This meant hiding in fear and living off whatever they could scavenge in the bush. Anyone unlucky enough to meet an RUF patrol was either killed or abducted.

The RUF occupied the area outside Freetown for six more months after ECOMOG drove them from Freetown. The remnants of the AFRC occupied the area known as Okra Hill until British forces attacked them. Around Rogbom, the rebels searched out the people hiding in the bush and told them to come back to the towns. Otherwise, they said, if they met people in the bush, they would kill them, assuming they were Kamajohs.

Pa Kamara is the headman of Rogbom. At first, Pa Kamara told me, the rebels talked to the villagers nicely, saying they were fighting for the people. Still, they made the villagers work for them: toting, farming, and fishing. One of the female teachers in the village told me that the women were sent to fish by baling water out of ponds. This meant standing all day in water up to their chests.

The rebels were always unpredictable, and at some point, the character of their occupation changed. According to Pa Kamara, they started killing people and "doing every other bad thing." A village a few miles from Rogbom was the real center of the killing in the area. Eighteen people were killed there, though no houses were destroyed. When the residents talk about the time the rebels were in Rogbom, they talk about the killing and the abductions, but they also talk about the slave labor (fishing, beating rice) and the hunger. One woman exclaimed to me, "At that time, we had no salt, no pepper, no Maggie (MSG), we had to eat that way!"

The rebels targeted Pa Kamara. He told me, "They tried to kill me, but couldn't." They shot him and cut him with machetes. Finally, they put a machete in the fire and heated it red hot in order to burn him. I saw the marks on his back and chest. "I almost died," he told me. "It was only God that saved me." His survival, possibly through magical means, impressed the people in his village, ensuring his political future. Pa Kamara said that at one point the rebels threatened to amputate his hand. They had his hand on the block and the machete in the air. His children were crying and begging for them not to cut their father's hand. One of the rebels saw Pa Kamara's ten-year-old son, Pa Sorie, and said he wanted him. In all, two of his children were abducted. In both cases the rebel who took the child "asked" for the child. Of course, he had no choice but to hand them over.[1]

Unlike Pujehun and Masakane, in this area there were no Kamajohs and no Gbethis. Some say that is why the rebels were able to occupy the place for so long: no one stood up to them. Pa Kamara seems glad that now they do not have to deal with a class of violent men in their midst. Certainly, being so close to Freetown the people probably thought they were safe up until right before they were invaded. On the other hand, being close to Freetown meant they were always close to the SLA and saw the "sobel" behavior of the SLA all along. Although Rogbom is a Temne-speaking village, my host there was raised in a Mende-speaking area. She told me that she could hear the SLA soldiers at the checkpoint saying in Mende, "If I had a gun, I'd go and loot in those villages." This made it hard for the villagers to organize any resistance.

The main reason I worked in Rogbom is the nature of the child soldier population there. Rebels abducted all of the children of at least a certain age after the January 1999 invasion of Freetown, and after a year and a half, all of the children had come back. That is to say, the child soldiers in Rogbom were only gone for a short time, and it was clear to the population that they had been forced to fight. There were girls as well as boys in the population. Also interesting was the fact that about half of the children had "spontaneously reintegrated," that is, they had escaped and come back on their own with no government or NGO assistance.

By the time of my residence in 2001, all of the children who had been abducted had come home. The main issues on people's minds were rebuilding destroyed buildings and rehabilitating farmland. Being close to Freetown, they were also close to the center of NGO operation and they were understandably strategizing about how to attract the most funding for rehabilitation projects. Of particular interest to me was the school rebuilding project. Rebels had completely destroyed the primary school in the village (a school that also served several nearby villages). The villagers had built a temporary replacement out of sticks and UNHCR tarpaulins. Although almost all of the children in the village had been carried away by rebels, only about half of them were formally registered with child protection NGOs as former child soldiers.

There was UNICEF support for rebuilding the school in proportion to the number of *formally* recognized ex child soldiers. My choice of field site was in part to allow a comparison of the experiences of the Rogbom people, who had formally recognized child soldiers in their midst, with the experiences of Masakane people, who had suffered just as much or more but did not have the same formal recognition of their children's participation in the war.

*   *   *

The preceding chapter was about the institutions that make up the formal rehabilitation and reintegration programs for child soldiers and to some extent create the identity "child soldier" as something with strategic value in postwar Sierra Leone. This chapter takes place outside the child protection organizations, back "home" in "normal" life. Serious strategic questions shape decisions by combatants: Shall I "be" an "adult" or a "child"? Shall I go to an ICC or reintegrate informally? What are mere bureaucratic classification issues for NGOs are a world of political maneuver for ex-combatants. In this chapter I address the whole population of children affected by the war: those reintegrated through formal channels, and those who made it home on their own—what NGO discourse would call the informal, or spontaneous, integrators. Finally, this chapter is about what happens in communities after children come home, and the complex and contradictory role that child protection NGOs play in postwar reintegration.

Here, the narrative follows my movement in the field from ICCs to villages. In ICCs, there was much speculation about the so-called spontaneous reintegrators, but in general those embedded in the system were only familiar with the children moving through their centers. For me, understanding the whole population of children affiliated with the fighting forces required moving out of ICCs and into the four sites described in the introduction: Pujehun, Masakane, Rogbom, and Jerihun.

Generally speaking, war-affected children everywhere were doing many of the same things as the children in the formal programs: attending school, participating in skills training, living in foster care arrangements, raising babies, and so on. The interesting questions became:

What are the effects of the bureaucratic distinction between formal and informal reintegrators? How do the practices and policies of NGOs affect the powerful political identity "child soldier," the children who were soldiers, and Sierra Leone in general?

Informal Reintegrators as a Problem

There are a large number of children who were child soldiers (according to the formal definition) who did not go through the formal demobilization and reintegration process.[2] "Spontaneous reintegrator" and "informal reintegrator" are NGO terms, used to describe the whole set of children who had been affiliated with the fighting forces but who did not go through the formal system. In a way, it is a residual category, defined abstractly as an object of NGO interventions not yet acted upon. A good estimate of the total number of informal reintegrators is, for obvious reasons, hard to come by, but it is almost certain there were more informal than formal reintegrators in Sierra Leone.[3] UNICEF and international child protection NGOs frame this large population of informal reintegrators as a problem, and phrase the problem in terms of children falling through the cracks of their interventions. UNICEF- and NGO-funded studies of child soldiers for the most part miss this population, since they are usually most concerned with evaluating their own programs and therefore their "sample" is the formal reintegrators.[4]

Many child protection workers have never met one of these "informal reintegrators"; nevertheless, a kind of conventional wisdom about informal reintegrators had developed within the system. For example, it is known that there were many girls involved in the RUF, but very few who demobilized. UNICEF reports that 5 percent of children who demobilized were girls (Brooks 2005), and yet interviews with ex-combatants lead us to believe that approximately one-third to one-half of the RUF fighting force was female.[5] Therefore, we can conclude that most girls were informal reintegrators. Similarly, we know there were many children in the CDF, but very few who went through formal demobilization. This creates a sort of top-down definition of the population or problem. That is, we start with reports that there were many more children involved in the fighting than showed up in formal programs, and then extrapolate from there.

In contrast to this deductive reasoning approach, I worked from the bottom up. That is, I started in a few locations (carefully selected to represent a certain range of wartime experience) and extrapolated up about the issues of informal reintegrators. Starting from the village level means that from one location, I saw a whole range of possible postwar trajectories for children. The purposive sampling is important: as described in the introduction, I chose distinct locations that I knew ahead of time would yield examples of the kinds of children who were not on UNICEF's radar: that is, informal reintegrators. Naturally, as an ethnographer, I could only work in a few places, and therefore the total number of children I worked with is a small fraction of the total. The top-down approach studies a larger number of children, but there is a selection effect. I studied a smaller number, but a wider range, of children. Neither one of these approaches alone yields the whole story, but together they can come much closer to understanding the total situation.[6]

UNICEF often claims that informal reintegrators are hard to find, but I found them very easily in every location I worked.[7] In fact, I met informal reintegrators everywhere I went. I made contact with many of them in schools I was observing near Freetown. My friend Wusu's brother introduced me to two boys in his class who had been with the RUF but had found people to sponsor their educations in Freetown. I became quite close to two boys who had fled the ICC in Lakka to make their way as driver's apprentices in Lumley. Finding girl informal reintegrators was a bit more difficult, and I will report more on the reasons for that in chapter 5. However, even after I stopped spending time in ICCs, I still found former child soldiers everywhere.

Informal Reintegrators as a Success

Although the primary focus of this book is not the evaluation of child protection programming, the obvious question about informal and formal reintegrators is "who reintegrates better"? Do the ICCs' interventions work? Or, perhaps more subtly, what are the comparative benefits of formal and informal reintegration? In a somewhat methodologically suspect study, political scientists Macartan Humphreys and Jeremy Weinstein report on the results of their survey of 1,043 ex-combatants (including women and children):

[N]on-participants in DDR did not fare any worse in reintegration than those who disarmed and participated in DDR training programs. . . . In fact there is evidence that among those that had a problem gaining acceptance from their communities, those that did not take part in DDR actually resolved these problems more quickly than those that did (2004, 45).[8]

To understand why this might be the case, let's start with a few examples regarding child ex-combatants from my field sites. Recall that in Rogbom, the entire population of children had been abducted, were away for at most a year, and then the entire population came back (either formally or informally). In this instance, reintegration was fairly easy, by which I mean that everyone in the village accepted that the children had been taken against their will, and gratefully welcomed them back home. This is not to say that children in Rogbom did not have problems: the school had been destroyed, and many did not have adequate shelter or food, but these were problems they shared with the community as a whole. Certainly, some children in Rogbom had a harder time returning to "normal" life than others, but overall, these were successful reintegrations.

Compare this example to the case of Abu Sesay, a formal reintegrator whose reintegration was not successful. I met Abu one day while riding on the back of a motorcycle with Alpha, a local child protection officer. We stopped abruptly at a small village and Alpha pulled his bike up to a large cement house. Children and women ran up, laughing and smiling. (As usual, people were surprised to see a white person on the back of the bike.) Alpha told me the village was named Masesay (meaning "the Sesay place"), and the boy he had reintegrated there was named Abu Sesay. Masesay was a very small village on the main road between Masiaka and Mile 91, made up almost entirely of members of Abu's extended family. Abu had been abducted into the RUF at the age of about thirteen when he was visiting an aunt in another town. He was the only boy in the immediate area who had been abducted by the RUF. This meant that people in the surrounding villages were wary of him. In fact, he could not leave his village, even to collect firewood, without an escort of his family members for fear of being attacked by people from neighboring villages. He was occupying himself with construction work in the village, but what he really wanted, he told me, was to attend school. There was no functioning school in the area however. Both Abu and members of his family told me that they were glad he was back with them, despite the many problems with his reintegration. When I asked why, they looked at me blankly and said, "He's family."

As we pulled away from Masesay, Alpha told me that was the third and last follow-up visit he would make with Abu. He acknowledged problems with the situation, but he had a large caseload of other children to visit.

Several months later, I was passing by Masesay with another child protection worker and stopped to say hello. I found out that Abu had left a week earlier, driven away by the lack of opportunities in his home village and lured by the prospect of quick wealth in the diamond-mining region of the east. A big man had come around, recruiting youth for his mining operation. So, although Abu had been counted as a successful reintegration for the UNICEF statistics, he was back with the "loose molecules," the so-called *san san* boys of the diamond fields, many of them his fellow ex-combatants.[9] This, then, was an unsuccessful reintegration.

Why Reintegrate Informally?

Why might children choose informal reintegration? This question requires a brief treatment of demobilization procedures, and demobilization as a particularly important juncture at which the formal/informal distinction was made. Demobilization (or "demob") refers to the formal process of handing over one's gun and leaving the fighting forces. Both adults and children demobilize. The demobilization activities in Sierra Leone after 1999 were handled by the UN peacekeepers (UNAMSIL). UN staff went anywhere there were a number of fighters ready to stop fighting, and set up a demobilization center. At the center they collected guns and ammunition and registered ex-combatants for benefits, assigning each a number.[10] The benefits were administered by NCDDR, the National Commission for Disarmament, Demobilization, and Reintegration. There were two parallel programs, one for adults and one for children. Adults were sent to a DDR camp where they received some training, some counseling, and a three-hundred-dollar resettlement allowance. They often brought their wives and children—some of whom could be categorized as child soldiers in their own right—along with them to the camp as "camp followers."[11]

Following international guidelines, children were defined as any person who was under eighteen at the time of demobilization. (The length of the conflict was such that the majority of children recruited, particularly to the RUF, were adults at the time of any formal demobilization process.) After demobilization children were supposed to go to ICCs, run by various child protection agencies, where they would be cared for until they could be reunited with their families. The desire to get them away from their adult commanders as quickly as possible was part of the design. Unlike the adult program, children who were not eligible for disarmament—those who showed up at the center without a gun—were not rejected but were sent to interim care for family tracing.[12]

According to UNICEF officials, when the first combatants started to arrive in the demobilization camps, it became obvious how distorted their perception of the process was. Children especially felt betrayed by a program they believed would provide them with three hundred dollars, immediate enrollment in school, vocational training, or access to employment. Some said their commanders had told them their reintegration benefits would include a Walkman and sunglasses. Some commanders had promised the children that in exchange for their demobilization allowance they would buy them clothing and "treats." The outcome was to create a level of expectation that could not be fulfilled.

Particularly for older youth, the adult demobilization allowance was a big draw. In some cases children tracked into the child program returned to the bush to find a weapon and attempted to qualify for the adult program. The information they carried with them about noncash benefits for children acted as a disincentive for the release of children.[13]

Susan McKay, a psychologist and women's studies professor, and researcher Dyan Mazurana found that although officially those under eighteen years of age were not required to present a weapon to enter DDR, there was widespread discrepancy among UN and NCDDR officials and staff of NGOs working with the DDR process as to whether or not children had to turn over a weapon (2004, 98). Furthermore, according to nearly all their respondents who passed through DDR, the weapons test with an AK-47 was repeatedly administered to children to determine whether they would be admitted into the program, that is, unless they could show that they knew how to "cock and load" they were not registered as ex-combatants (McKay and Mazurana 2004, 100).

These misunderstandings of policy led to strategic decisions on the part of ex-combatants about how to position themselves. At the national Child Protection Committee[14] meetings I heard examples of fifteen-year-old "adults" and twenty-year-old "children." There were no birth certificates, and no clear-cut way to make the distinction between under-eighteens and over- eighteens. In practice, a certain set of combatants could conceivably portray themselves either as adults or as children. An individual preparing to demobilize had to weigh the two different tracks based on future aspirations, personal history of school attendance, and an assessment of which promises were more likely to be kept.

How were formal distinctions made at demobilization about who was an adult and who was a child? This task fell to the international UNAMSIL staff, often overworked and underprepared, and perhaps easily swayed by the protestations of those demobilizing. I heard that one method UN staff used to determine who was a child was whether wisdom teeth had come in. There was grumbling among the child protection staff that Sierra Leoneans would be better equipped than foreigners to make the distinction (pointing again to the fact that childhood is a culturally variable construct). This distinction making—who is a child, who is an adult; who is a soldier, who is not—is the first step in making, in a way, "child soldier." As soon as it must be recorded or enumerated who is a child soldier and who is not, the line is drawn and takes on a bureaucratic reality of its own.[15]

Perhaps the children did not choose whether to go the formal or the informal route but were cheated out of reintegration benefits by their commanders. Paul Richards and colleagues (2003) report on ex-combatants who claim to have been defrauded by their commanders in the demobilization process and as a result have been denied reintegration assistance.[16] Although officially children were not required to present a weapon to demobilize, there was a great deal of misunderstanding on that point, so many children may have thought they were ineligible for DDR.[17] Perhaps they did not even know that formal demobilization was an option.

Macartan Humphreys and Jeremy Weinstein report on their survey that, "significant numbers complained that they or members of their communities were not able to gain access to the DDR process at all" (2004, 3).

More importantly, perhaps combatants (or their family) wanted to hide or forget their participation in the forces.[18] Forgetting instead of remembering is a key strategy for dealing with trauma, and the strategy of secrecy is at the heart of understanding why some chose to spontaneously reintegrate. Michael Jackson offers an explanation of why one might choose to go the informal route. It is tied to the idea that forgetting might be better than remembering as a way of dealing with trauma. He recounts the story of a man whose son and daughter had been abducted by the RUF. The son managed to escape during a battle to dislodge the Sierra Leone Army from Makeni, and he returned home to Freetown and his family. His father told him that he did not want to hear anything of what happened. It made him feel bad. As for the boy, apart from saying he hated the RUF and would never forgive them for what they had put him through, he wanted only that his ignominy not become public knowledge. During the disarmament period, his father urged him to go and find his weapon and hand it in to the authorities, but his son said, "No, I want no record of the fact that I carried arms; I will not do it, even if I am paid millions of leones" (Jackson 2004, 72).

Informal Reintegrators in Rogbom

Although generally I found Sierra Leoneans amazingly willing to talk to me about the war and their roles in it, sometimes people were reluctant to talk. Rogbom was small enough (about twenty households) that I was able to talk to almost everyone there. The hardest interview was with the mother of two boys, Morlai and Mohamed, both of whom had been abducted. She sold cookery (prepared rice and sauce) along the main road. She did not seem particularly anxious to talk to me and answered my Krio questions in Temne, though my host told me she could speak Krio and she understood the questions I asked in Krio. Her reluctance is understandable based on her experience. Her son Mohamed was kept in the Pademba Road Prison in Freetown for four months after the events of May 2000 when RUF rebels captured and held hundreds of UN peacekeepers for a few weeks. Although he was a registered ex-combatant, living in the city and receiving skills training, he was one of the many former RUF rounded up and jailed at that time. His mother had understandably lost some confidence in the demobilization and reintegration system for ex-combatants. She also talked about how she spent Le 4,000 (about $2, but still a day's earnings) for round-trip fare to Freetown for Morlai, the younger boy, and three days in Freetown so he could "register," and she has yet to see any benefit from it.

Kadiatu, a girl of around fifteen also from Rogbom, was an informal reintegrator. She said she had been taken away and lived with the rebels for about eight months. Eventually, she and some other girls were taken to live at Waterloo (near Freetown, and only a few miles from Rogbom) with the commander's auntie. Someone from Rogbom saw her at the market in Waterloo and came back to tell the family where she was. Her uncle went to claim her back. When the uncle arrived at the house where she was staying, the woman he met there said that since the woman who had Kadiatu was not there, she could not turn her over. When he asked the girl why she had not returned on her own, she said she did not have the money for transportation. After the uncle returned to Rogbom empty-handed, everyone in the village talked to him and said he should have come with her anyway. After a week, he went and took her. She told me she had to leave her clothes behind, but she is much happier to be home.

Back in Rogbom, there were complaints about her behavior, that she no longer *yehri wohd* (obeys, literally "hears words"). When I asked her about it, she admitted she was not used to following orders, but that she was coming around. I asked her uncle if he ever talked to her about what happened in the bush, in particular if anyone ever "married" her. He said he never asked her about it: "It's better not to think about it." She said she sometimes thought about that life. She is glad to be in school now, though sometimes her companions provoke her and the others. She still sees other children she knew in the bush and even some of the rebels/AFRC, who, after their own formal DDR, have now joined the new SLA and gone through the British training at a camp nearby.

How Communities Do Reintegration

There is a Krio saying: *bad bush no dae for trowe bad pikin* (there is no place to throw away bad children). It means that even if children are bad, they must be tolerated, and taken care of. It is this belief that lies behind the success of community-based reintegration. To say a child has reintegrated, I heard people say (in Kringlish):

> *In at don kol* [His (or her) heart has gotten cold[19]]
> *I dae kam tu small small* [He (or she) is coming around little by little]
> *I dae take control* [He (or she) is submitting to control]
> *I don dae change* [He (or she) has begun to change]

The opposite:

> *I dae wild* [He (or she) is behaving wildly)]
> *I no dae take control* [He (or she) will not submit to control)
> *I no dae yehri wohd* (He (or she) is disobedient—literally, does not hear words]
> *Da rebel style still dae pan am* [That rebel style is still upon him (or her)]
> *I radical* [He (or she) does not respect hierarchy]
> *I crack head* [He (or she) is crazy]

Kadiatu's reintegration was happening, as her family told me, *small small* (little by little), through her daily interactions and mixing with the community and her peers. What seemed to be most useful for children was reinsertion into normal life. This meant attending the same schools as their peers, and so on. It was participation in everyday activities that seemed to bring children around. And, in most cases, the informal reintegrators seemed better able than the formal reintegrators to return to "normal" prewar life and reintegrate into "normal" Sierra Leonean childhoods.

Informal reintegration happened through many of the same practices discussed in chapter 2, but with a different flavor. As the following examples show, informal reintegrators often took advantage of fosterage or apprenticeship, which made for longer-term solutions than short-term ICC-based interventions.

Sheku was an informal reintegrator I got to know fairly well. He is a good example of how fosterage led to informal reintegration. He was a student at the secondary school near my house and was in the same class as Wusu's brother Joseph. I interviewed Sheku at my house after Joseph brought him around saying, "Aren't you interested in child soldiers? My friend Sheku was one." Sheku told me that he had been in the bush with his uncle and brother collecting palm kernels when they were unlucky enough to meet a rebel group. The rebels killed his uncle and took him and his brother captive. He told me the story of his abduction and training by the rebels. He was eventually released and was now living with a family in Freetown. He was anxious to continue his schooling after missing several years.

I visited Sheku's house in a suburb of Freetown after interviewing him. I met his guardian, Ahmed. Ahmed told me that he used to see Sheku around Kono (in the east) before the war while he was working for the Ministry of Agriculture. Sheku's parents had fled to Guinea, so Ahmed took him on as his boy, though he said, "He's not like a boy. He's more like a little brother." He said in fact it was Sheku who saw him and came up to ask for help. So, they came to Freetown and Ahmed is paying for Sheku's school fees and feeding and Sheku *bruks* (washes clothes) and does other work for Ahmed. Ahmed is living in Freetown with his mother, his wife, and some other dependents. He says Sheku is doing very well in school and is a great help to him. However, Ahmed is planning to go to Ghana for a year and a half for further training and he has no way to take care of Sheku. He asked whether I would be able to help him get Sheku included at the ICC at Lakka so at least he could stay there for a year or so and they could pay his fees. I demurred, saying the goal of ICCs was to move kids along by reintegrating them with their families, and they did not want to keep them that long.

Neither Ahmed nor Sheku seem to be in any rush to find his real parents, and they never discuss his ex-combatant status. Rather, their arrangement is a pragmatic taking on of a boy in the tradition of child fosterage. There is a real difference between these people and those fully plugged into the system. There is no discussion of being traumatized, for example. The only concern is that the boy should be in school.

Apprenticeship is also an important method of informal reintegration. There are many child ex-combatants who never went back to their villages or families of origin and instead found a living on the streets of Freetown (or other urban centers.) These are the apprentices and street children. Some of them I would consider successful reintegrators. I found two ex-combatant boys from Lakka who were apprentices to a driver, or rather they found me. The two boys remembered seeing me at Lakka while they were still living there and were excited to have me come and meet their new caretaker. The driver's name was Victor, and he and the boys slept in his vehicle that they parked near the petrol station in Lumley (the main transit point between the center of Freetown and the Lakka ICC). Victor explained that the boys were sometimes difficult, always fighting and wrestling, but he felt he had to take them on. The boys said they were not getting the promised support from FHM, so decided to strike out on their own.

I visited Victor and the boys many times over the year when I was transiting through the Lumley area. Victor was not doing that well for himself, and the two boys eventually turned into general-purpose boys in the area, carrying loads for people and so on. I gave them a little money every now and then when they asked for it. They never seemed to want to go back to Lakka. These boys became informal reintegrators, because they left the formal system. They never went back to their families. Was theirs a successful reintegration? They were living a life of poverty, but so were most kids in the area. They had a guardian of sorts, they were making a living, and their ex-combatant status was never an issue. Maybe they should have gone back to their families, but they seemed to be doing what they wanted to do.

Although "informal reintegrator" is, for NGOs, a signifier for the unknown, an undifferentiated mass, for children actually making their way in postwar Sierra Leone, reintegration is partially determined by many factors: among them gender, location, level of education, and fighting faction. Recall that I chose field sites specifically to try to get at these very social and political divisions and distinctions, and in this chapter so far I have looked at examples from Rogbom and Freetown. I next turn to Pujehun, the large town in the south with the longest experience of conflict.

Pujehun is a district headquarter town in the Southern Province with a population around 10,000. It looks like an average Sierra Leonean town with mud-block or cement-block houses, with mostly zinc pan roofs, and dirt roads. Pujehun means "pepper place" in Mende, probably referring to a history of pepper cultivation or trade. At the time of my fieldwork, in juxtaposition to the massive destruction surrounding it, there was a well-functioning hospital in town, recently rehabilitated by the NGO Médecins sans Frontières (Doctors without Borders). The Catholic Church has been in Pujehun since 1912 and there is a big Catholic compound. There are two large Catholic secondary schools (one for girls and one for boys) and seven primary schools. There are district and local offices and a prison, and market stalls surrounding the hub of public transportation and commerce. There were even two bars selling cold beer.

I chose to work in Pujehun for several reasons. I knew that as one of the first areas impacted by the war, I would find there boys who had been recruited by the RUF many years ago and therefore also had been reintegrated for the longest time.[20] In addition, the local Catholic-affiliated child protection NGO, Children Associated with War (CAW), was active in Pujehun. It was one of the first NGOs to work with child soldiers, and, because it was local, it had been doing this work even before the dominance of UNICEF. It had fallen on hard times since the influx of international NGOs.

The war has a long history here that most trace back to the violent events following the 1983 national elections. My host in Pujehun explained to me that in 1983 a local election was rigged in favor of the APC's favored candidate, and after the election a group of people joined to protest and killed supporters of the winner and burned their houses. The protest came to be known as *Ndorgbowusu* after a local "devil" supposed to have made the protesters invisible while they carried out their attacks (Kandeh 2002, 188).

In 1991, when RUF rebels entered the country from Liberia, there was initially some support for their political program in the south.[21] Some of the earliest RUF recruits were from the Pujehun district. For many years, the war was fought only in the south and the east (Mende and Kono country), both because of political opposition to the APC that grew out of ethnic alliances and also because the south and the east are the greatest diamond-producing areas. The complaint had long been that although the south and east (Mende and Kono areas primarily) produced the greatest wealth, the government was run by northerners (Temne and Limba).[22]

My hosts in Pujehun were two CAW workers who were native to the town, Sylvester and Sowa. They took me around to introduce me to the local chiefs and the head of the CDF in the town. During a walking tour of the surrounding villages, they showed me the exact spot where the rebels killed the first person in the area, a soldier on a bike.

In addition to being one of the first places the RUF came, Pujehun was also one of the first places that the CDF became strong. At the time of my fieldwork, the local CDF commander was very powerful and very involved in politics and making the CDF into a real political force. The Kamajohs certainly did some bad things in Pujehun: looting, extortion, and shooting. Still, people seemed to forgive them for it (though there was still some fear lest their power become too great.) They liberated Pujehun and were still in control there. I saw no UNAMSIL and no SLA troops.

I expected to meet children who had been demobilized four or five years previously, now getting on with their lives in school or in other activities. What I found was that almost all of the children who had been recruited or abducted by the RUF or SLA had been re-recruited into the local CDF as soon as they returned home. Some who had joined the fighting early on had never come back. Even at the time of my fieldwork, although the local CDF commander denied that they used child soldiers at all, I always found large numbers of "youths" hanging around the CDF headquarters. For political reasons, the CDF commander told me that 90 percent of male adults in Pujehun belonged to the CDF and that no children did. The CAW workers told me the numbers were probably more like 60 percent of adult males and 30 percent of children.

As a result of its specific war trajectory, there were some odd hybrid categories of reintegrator in Pujehun, such as what I call the semiformal reintegrator. Because the RUF came and went through Pujehun in the early 1990s, and CAW started before the formal DDR program, many of the children were demobilized before there were formal programs for them to participate in. CAW's original beneficiaries were "vulnerable children" from the Gondama IDP camp, with a few ex-RUF or ex-SLA children in the mix. Their focus was certainly not "child soldiers" because that term did not yet exist for them. Years later, most of their beneficiaries did not have official demobilization numbers, so they found themselves in a strange space bureaucratically.

To summarize: in Pujehun the population of "child soldiers" was made up of a large number of active CDF (although their presence in the force was denied by their leader), and a handful of ex-RUF. A few of them were in school, the rest were doing some kind of skills training. There was very little NGO help for child ex-combatants. The population of child soldiers was not well documented and not well organized.

At the time of my fieldwork, the CAW program was barely funded. However, despite their lack of funding, or perhaps because of it, they were quite successful. Sylvester, one of the the CAW workers in Puje-hun, tried explicitly to work within already existing institutions in the area. He was able to get some of the ex-Kamajoh boys apprenticeships with local masters, some as carpenters and some as tailors, without financial support from CAW.

It was not like a formal skills-training program; rather he met some of the boys in their villages and convinced them to come do *bete tin* ("better thing," or something worthwhile). The carpenter's apprentices seemed pleased with their new skills. I asked the master carpenter why he had taken them on as apprentices. He talked about the need to pass on his skills, and also said something about how everyone needs to help these kids. But again, I do not think it was so much that they were ex-combatants (that is, those poor damaged children) or ex-Kamajohs (that is, our protectors who need rewarding) as much as it was helping young men who had no other options, and needing apprentices anyway (much like Victor the driver discussed earlier.)

In summary, then, fostering and apprenticeship are two very important informal reintegration strategies. A list of possible informal reintegration trajectories would certainly include those child combatants who left the bush and went home, those who left the bush and became street kids, and those who left the bush and found foster parents or masters for apprenticeship. In addition, one would need to add the semiformal reintegrators: those who left formal programs and went home, into fosterage, or apprenticeship on their own. All told, that is quite a number of possible trajectories not really accounted for by the assumed life cycle of the child soldier.

There is the danger that I might be misinterpreted here, so I want to be very clear. I am not saying that children affiliated with the fighting forces are unaffected by their experiences. They have seen and participated in terrible acts and are sure to be affected for years to come, with a diversity of effects across individuals. However, in many ways, they are no more "traumatized" than anyone else in the community. There are superficial

Figure 4.1. Former Kamajohs in carpentry apprenticeship in Pujehun.

changes in their behavior, like differences in slang or style, but "the ones carried away" are generally no more troublesome than your average "troublesome" children in any village. One cannot assume that child soldiers are more traumatized than anyone else—it very much depends on their experiences and the experiences of the community. Perhaps more important than my perspective on this is the fact that Sierra Leonean villagers did not seem to see the ex-combatants as a particular problem, any more difficult to reintegrate than various other collaborators. This point may come down to forgiveness, and how Sierra Leoneans are choosing to move ahead in rebuilding the postwar nation. The main conclusion is that, for now, the problems of former child soldiers would be almost unremarkable in the village setting if not for the interest of international NGOs.

What We Mean by Reintegration: The Best Interests of the Child Revisited

So we return to the question, what is successful reintegration? Is it a return to a normal Sierra Leonean childhood, or does reintegration require (as was argued to me by UNICEF staff and researchers) certain standards of care? Dissatisfied youth were the tinder for the war, so in some ways it is dangerous to just put them back the way they were. The question for child protection professionals, and for postwar Sierra Leonean society in general, is this: Is the goal to reinsert children, to the extent possible given the destruction of the war, into the traditional power structure, or is it to try to change the power structure? This is obviously—at least to me—a *political* distinction, with important ramifications. Are child protection agencies concerned with changing political structures or with guaranteeing the welfare of individual children? Indeed, this is the unspoken problematic behind DDR programs in general.

To address this question, we must return to an earlier question: who reintegrates better? The answer is that if you prefer the reintegration into normal life, then informal reintegration is the most effective; if you prefer to "improve" the social position of children, then formal reintegration is the answer. In fact, it is not a question of reintegrating "better" or "worse," but of two different kinds of experiences with two different kinds of outcomes, for individuals and for society.

What is the net benefit of spontaneous reintegration as opposed to formal reintegration? The community has fewer expectations when a child comes home on his or her own, and the child can more easily reenter community life.[23] The returning child can make use of the strategy of keeping the details of his or her abduction, participation in fighting, and other possibly embarrassing facts secret. The negative side of informal reintegration is that some children miss out on some benefits, including medical and psychological screening, so the few really traumatized cases do not get the help they need.

The formal reintegrators certainly have more knowledge of the world of NGOs and of the strategic uses of Western discourse about "the child." That is, they can make better use of discourses of abdicated responsibility, and this in some ways eases their reintegration. On the other hand, although they got more support (for example, medical and material aid), they had to adopt a new identity with respect to their communities. That is, they had to reveal themselves and their war histories, and therefore sometimes became the focus of anger at the inequitable distribution of resources. Informal reintegrators often did better at reintegrating, though formal reintegrators got more support.

Informal to Formal: "How Do I Get My Name on the List?"

This book is concerned with the identity "child soldier," not just for what that identity means to the individuals themselves, but, in some ways more importantly, as a social category. The questions become: How is the distinction between formal and informal reintegrators made and unmade, and to what effect? What are the practical uses of the construct "child soldier" in Sierra Leone? This section explores how individuals and communities struggle for the label "child soldier" after the fact of reunification.

Often it was only after reintegration that children realized how valuable their "formal" status could be. The boys I met at the Jerihun camp showed me the "bangles" (plastic bracelets) from their demobilization that proved they had been through DDR. They told me the bangles were among their most prized possessions. I saw this phenomenon often. Another example is from Gbado, a village I was considering for fieldwork but eventually had to decide against because UNAMSIL felt it was unsafe. The one day I was there, meeting the headmen of the village and getting the tour, a boy came up to me, an unknown white lady, and unbidden showed me his bangles, saying, "If there is any benefit to come, don't forget me."

Some early interventions on behalf of child soldiers made the distinction between formal and informal reintegrators painfully clear. For example, the Canadian International Development Agency (CIDA) sponsored a program to give all formal reintegrators new school uniforms, the idea being that this would ease their reintegration back into schools. However, it turned out that in most communities struggling to return to normalcy after the war, almost none of the children could afford school uniforms. The end result of the program was to make the few registered child soldiers stand out among their peers. That did not help them blend in, and it provoked resentment among the other students. There were naturally some problems with the administration of the program, with children getting the wrong-size uniforms or uniforms for the wrong school. I spent a day with a child protection worker, riding from school to school in his operational area, handing out uniforms. When his charges complained about getting the wrong uniforms, he advised them to try to find some other student to sell the uniform to. When other children complained that they would like new uniforms as well, he advised them to be patient and wait for some other aid program to come along.

The glaring problems with this program, and other similar programs, led UNICEF and its implementing partners to come up with a different program entitled the Community Education Investment Programme (CEIP). The program was "aimed at building the capacity of schools to meet the learning needs of all their students through assistance given for the admittance of demobilised child ex-combatants and other separated children" (field notes). In other words, for every formally demobilized ex child-soldier a school admitted, it would receive help that could be used by the school as a whole. In exchange for all fees and charges being waived for one registered child, the school got one package from the choices presented in Table 4.1.

The supplies were to be overseen not by the schoolmasters (often seen to be corrupt) but by the local Community Teacher Association/ Parent Teacher Association (CTA/PTA). The CTA/PTA was instructed to "ensure that any supplies the school receives are distributed equitably and have a direct benefit for as many children as possible in the school" (field notes). The CTA/PTA was required to meet regularly, have free and fair elections, and follow other specific guidelines.

This program was a step in the right direction, but it was not without its own problems. How much aid a school would get was, sadly, basically a random decision, depending on how many formal reintegrators happened to live in the school's enrollment area. I saw the administration of the program up close while I was resident in Rogbom. The school in nearby Mayombo was doing really well, but only got one CEIP package because it only had one formal ex-child soldier. Another nearby school, in Mathiri, was barely functioning, but, because of the higher number of formal ex-combatants enrolled at the school, got about ten packages. It does not seem likely that they were able to make a functioning school with just the ten packages. More likely, the teachers and others sold off the supplies.

In Rogbom, more than half of the children who were "the ones they carried away" were not registered as ex-combatants and therefore did not get benefits. There was a lot of interest from community and school leaders in how to get them signed up, but not a good flow of information. Michael, the Caritas worker who introduced me to the village, said he had among his caseload six registered children in the school, though

Table 4.1. Package Contents

Package 1	Package 2	Package 3
200 ruled exercise books	Chalk—ten boxes of 100 pieces each	Football, junior size—5
200 squared exercise books	Blackboard paint (one can—8)	Ball-inflating kit—4
400 blue pens	100 black pens (for teachers)	Volleyball net—1
400 pencils	100 red pens (for teachers)	Volleyball—2
		Whistle—4

I wrote down names of about eighteen ex-combatants—some from Lakka (an ICC administered by a different NGO, therefore their names were not on the Caritas rolls) and some that "came for themselves."

In Pujehun, the principal of St. Paul's Secondary School, a venerable old boys' school, showed me around. He told me he was in the process of piecing together funds for rebuilding the school.[24] A few ex-combatants were attending the school; I interviewed three of them. The boys complained to me that they had previously been driven from school for not paying school fees since CAW could not pay for them. The principal denied this and showed me the register with records of school fees paid. Some had "UAC" next to their names to show that they were sponsored by the Christian Brothers Unaccompanied Children Project (another child protection NGO operating in the area), but none had "CAW" next to their names. I asked him if he knew about the CEIP program.[25] I put it to him that he should not be driving anyone with a DDR number from school, and in fact he should make sure he had all the boys with DDR numbers included to maximize his benefits. He said they had three boys with DDR numbers and that IRC had brought only one package. He showed me the books and pens locked in his cabinet. Later, I went to check with the CAW workers and they agreed there were only three boys with DDR numbers, though they estimated there were twenty to thirty ex-combatants attending the school. (This is an even lower percentage of formal reintegrators than in Rogbom, where about half the former child soldiers were formal.)

This system, and others like it, led to informals trying to get reclassified as formals after the fact. Everyone knew there were a lot of informals out there, and NGOs also had something to gain by having a larger number of beneficiaries on their rolls. Community brokers, such as headmen, NGO workers, teachers, and headmasters knew that their communities could get more aid if they could show they had more child soldiers.[26]

The point is not that there should have been better-organized programs for distribution of benefits. Any kind of program would have problems and people trying to work the system. Rather, the issue is the effect of these distinctions and the sorts of struggles that can happen when the identity "child soldier" carries benefits with it.

On the other hand, some of the stigma of having to be marked publicly as a child soldier dissipates as it comes less and less to reflect reality, and to be more a way to work the system. So the distinction comes to have less meaning as more children (whether ex-combatants or not) are added to the lists, thereby undoing some of the distinctions described above.

Community Strategizing and NGOs

Part of what I had to do to set up residence in Rogbom was to visit the headmen of the village and the surrounding villages. At Mathiri, a village near Rogbom, the chief made an interesting speech on my arrival about how white men have clean hearts and black men are wicked. These speeches always made me uncomfortable, for obvious reasons. He noted that if any white man comes to help them, he always asks about the welfare of the women and children. (I fit right into that pattern.) I saw this as a public acknowledgement that they know that whites—that is, international NGOs—have different interests than they do, and that they must be ready to discuss the issues of women and children if need be.

The following excerpt from a meeting of a local child protection NGO will make the point more clearly. I attended the meeting of a locally organized NGO whose founders were trying to set up programs for war-affected children in their home district of Moyamba in the Southern Province, but also for war-affected children from Moyamba now living on the outskirts of Freetown. This meeting was arranged partly for my benefit and partly to give the head of the NGO an opportunity to explain to adults what they were planning. He started by telling the assembled people that he had heard that four or five NGOs had come around and taken names, that is, registered "war-affected children" as beneficiaries. He exclaimed angrily, "They ate the money and we did not get anything." He continued,

> We are bringing a program for children, but *if di pikin get Le 10,000, yusehf mohs it 1,000 leones de* [if the child gets Le 10,000, you must "eat" 1,000 leones of it]. Everyone will benefit from this program. Whatever the white men send, if I see something I want I'll take it of course. But we should make sure the children are satisfied.

I was surprised at this explicit public understanding—in front of a white person no less—that adults will "chop" from children's benefits. He continued, jokingly adding, "When registering orphans, parents tell their children, '*kill mi wan them, leh bete mit wi* [kill me off right now, so that we will all benefit]." This was an explicit public understanding that the list of beneficiaries is a fiction.

Clearly, it is not only individual children who strategically deploy the "child soldier" label. Communities organize their self-presentation around the idea of "war-affected youth" in order to gain access to a certain amount of international aid: money from UNICEF to rebuild schools that register child ex-combatants, micro-credit loans from the Catholic Church for families that foster child ex-combatants, and so on. Communities learn to "talk the talk" of child rights and cast their problems as problems of youth. Communities fight over how many ex-child soldiers they have, and they try to get more young people signed up. One activity repeated in many communities was the creation of a list of child soldiers, to be ready in case an NGO with ready funds for reintegration programs came around.[27] The lists were generally drawn up by the headman of a village with the help of the local school headmaster, sometimes formalized as the "village child protection committee."[28] Notice that the CEIP follows that model by calling on a CTA/PTA to administer the benefits. In one case I observed, it was the local Civil Defense Force commander who decided who should be put on the list. The lists thus compiled generally did not match what I had come to know about the actual participation of children in fighting. The chief's son, the imam's son, and those who were already attending school appeared on the list—all youth who were *not* former combatants.

Inclusion on the list, in this case, was based on connections, not necessarily on who had actually participated in fighting. But there was also something pragmatic about it. Everyone would benefit by claiming larger numbers, and by claiming students instead of nonstudents; children who did not attend school would not qualify for aid to schools.

I am not the only one to comment on this state of affairs. Researchers Paul Richards, Khadija Bah, and James Vincent discuss the phenomenon of "briefcase NGOs" in Sierra Leone and the general proliferation of NGOs under war-time conditions.[29] They conclude that "villagers often became quite adept at playing the agency game—knowing how to ask for what agencies had to give even when this was not a local priority" (2004, 26).

While I was in residence in Rogbom, a letter came for the headmaster with the names of local people who had been through DDR, asking them to come to Masiaka to receive agriculture inputs (seeds, hoes, fertilizer, and so on). It turns out that many of them had used false names, had gone elsewhere (mainly to Kono to dig diamonds), or had lost their ID cards. Some were regularly going elsewhere to work on other ex-combatant projects. The teachers and the chief all got together to discuss this problem. They wanted to get the biggest possible assistance package for the community and would therefore need to send all ten people. One of the teachers figured all they really needed was ten people in the group and that if one or two of those named were not *actual* ex-combatants, it would not be that bad. (The teachers have an important role to play translating the "white man world" for people to help them get access to benefits.) However, there was danger in this. The same NGO was also sponsoring an agricultural assistance program for ex-combatants in another nearby town and they had to make sure they did not duplicate any of the names of people already registered there.[30] The headmaster ended up sending one of the teachers (during school hours) to the agricultural project to find the names of those already participating, warning him not to show them the list or the participants would give a false name just to be included twice.

This kind of manipulation of official lists for political reasons is not new in Sierra Leone. These manipulations are as much reconfiguring old circuits of power as they are bringing new forms of power into being. Anthropologist Mariane Ferme writes about various counting exercises carried out by the state—censuses, taxes, elections—in areas of Southern Sierra Leone, and concludes:

> Given the ambiguity of the state's use of numbers—sometimes to benefit, other times to benefit *from* its citizens—many rural Sierra Leoneans saw counting and defining as contentious issues. To them, these were not technical procedures for neutrally recording statistical information to be used by a bureaucratic apparatus, but rather political acts aimed at exposing and controlling people (1998, 160).

My impression of life in Rogbom was that a main activity was to get funds for various types of projects. Since many Rogbom people went to the Approved School Camp for internally displaced persons in Freetown around the time that the rebels occupied their town, they had gained an understanding of the NGO supply system. In fact, some people's survival strategy was to live at the camp and come to Rogbom occasionally to see how things were progressing. There was a lot of work going on in Rogbom, such as sawing boards, making palm oil, and harvesting mangoes. Still, there was no seed rice (they were waiting for an NGO), no roofing zinc (they were waiting for an NGO), and the school reconstruction was supposed to come from the Government of Sierra Leone[31] and supplies such as blackboards and desks from UNICEF.

Near the end of my time in Rogbom, the local Caritas Makeni worker called a meeting of the zonal Child Protection Committee (CPC). One of Caritas's goals at the time was to set up local child protection committees in every community in which it worked, so that local communities would "have ownership" and take child protection seriously, but also so that there would be a community participation structure to show donors.[32] The committees were usually made up of teachers, headmen, and other important and literate people. The point of the zonal meeting was to get all the local CPC members from around the area together so Caritas could explain its benefits package to everyone at once.

The headmaster came to the headman about the meeting. He explained that Caritas had asked the village to provide Le 5,000 (about $2.50), two pints of palm oil, and twelve cups of rice, so they could cook for the meeting. The headman called upon everyone in the village and asked them to *hib* (contribute) five "block" (Le 500 or about 25 cents) each. There was some grumbling from the villagers, and I could not really blame them. Why should they give money to get a benefit they have not seen yet and they do not understand? Also, in the headman's description no mention was made of "reintegration of child ex-combatants" or even of children.[33] Participation in the meeting was sold to the community as necessary for bringing some unknown benefit in the future. On the other hand, the headmaster and his group complained to me about the community: "They never want to do anything, they only become interested when the supply truck pulls up, if they do not understand they should leave it to those of us who do understand."

I tagged along to the zonal CPC meeting in Mamamah (I decided to pay for our whole Rogbom contingent to go there by bush taxi rather than walk five miles). There was one Caritas worker present at the meeting to explain the program. One woman got up at the meeting and said, "Help us. We've all suffered a lot. We've been disappointed by other NGOs a lot. We hope you will help us through the children." The headmaster from Rogbom chaired the meeting. He tried to convince people that Caritas was a good investment of their time (and money). He said,

> Remember PLAN?[34] At first we thought *na plan den get for wi* [they had a bad plan for us]. They did not sign up all the children, but they brought "sibling" benefits. ... "A sibling is a person who is connected to the foster child whom God has brought light to." ... Caritas is another good agency that works like PLAN. It's not a lie-lie organization.

Again, the message was, do not worry about the specifics of who the beneficiaries are meant to be. If we play our cards right, everyone will benefit from this program.

* * *

Are NGO actions changing Sierra Leonean conceptions/practices of youth, and thus the very structure of society? In Rogbom, at least, there seems to be little impact. People line up and sign up for whatever program comes along, changing identities (even names—witness the DDR list) to suit the occasion. Meanwhile, life goes on as before, except that things are much tougher after the war, and the people's focus is on building houses, restarting agriculture, and so on. They will take whatever comes, I think, because they know they can bend it to meet their own needs.

Conclusion

This chapter started with the hypothesis that the study of informal reintegrators alongside formal reintegrators might be a good way to evaluate the effects of the formal programs. In the case of individual child ex-combatants, informal reintegrators certainly miss out on some NGO-sponsored benefits, but the difference between formal and informal reintegrators in terms of eventual reintegration into communities seems minimal. In a way, the processes I have described in this chapter level the playing field for child ex-combatants—when any child can be registered, no one is stigmatized. This state of affairs eases some things, and makes some things more difficult. It addresses the issue of child soldiers being singled out in their communities, since when a full range of children—ex-combatants and not—are registered, the label "child soldier" loses some of its sting. However, the wholesale acceptance of modern discourses of youth innocence may also disempower children. By adopting the modern notion of youth, young people gain one type of power and lose another. Children move from power that comes from the threat of violent response to injustice or inequity to a power legitimated through international structures, one that requires them to take on certain modern identities. The construction of children as innocent makes them silent and apolitical, and about potential rather than actuality. That is, children are important because they are the future of the nation, not because they are political actors in the present.

What are the results for the society as a whole? The cynical answer is: there is a new skill at artifice. But this kind of artifice is not new; it is simply framed now by the issue du jour, war-affected youth. A more hopeful question might be: is there in patriarchal Sierra Leone a stronger concern for children than before the war? I believe that remains to be seen. Children are certainly seen as valuable resources in a new way. It cannot be denied that the rise of NGO activity around child rights allows for the provision of much needed material resources for reconstruction. However, the reality of its distribution may undercut the well-being of children and youth. That is, youth continue in their position in the gerontocracy in order to gain access to resources that are distributed unequally by the elders. The provision of aid for former child soldiers is an example of how humanitarian aid can buttress patrimonialism in local communities. As Steven Archibald and Paul Richards conclude, "[F]ar from 'teaching' people their rights (as has been alleged) humanitarian activity . . . provided the resources for a modest renewal of patrimonialism" (2002, 358).

Perhaps most interesting, how is power/knowledge taken up and used strategically by the targets it seeks to govern? This chapter describes some of the complex and contradictory ways conceptions of the "child soldier" are implicated in the process of postwar reintegration of child ex-combatants. New meanings of youth as a political identity are emerging in Sierra Leone, meanings influenced by international discourse but resulting from the actions and agency of local community members and child soldiers who engage with the process of national reconstruction. "Child soldier" as a category is cocreated by both sides in social practice. Struggles over childhood and child rights in postwar Sierra Leone are productive sites in that they are becoming the locus for all kinds of other political struggles.

The next step, therefore, is to map the space of the deployment of the category "child soldier," that is, to examine the differential power of the social construct across time, region, ethnicity, faction, gender, and so on. How, where, and why does "child soldier" gain purchase in practice? The chapter that follows discusses this with a focus on two very important distinctions: fighting faction, and gender. In essence, the chapter first shows how former RUF and former CDF child combatants have differential access to the identity "child soldier" (and all that it entails). It then discusses how boy and girl ex-combatants have differential access to the category "child soldier."

5

Distinctions in the Population of "Child Soldiers"

RUF and CDF, Boys and Girls

At any particular moment, in any marked event, a meaning
or a social arrangement may appear free floating, undeter-
mined, ambiguous. But it is often the very attempt to harness
that indeterminacy, the seemingly unfixed signifier, that ani-
mates both the exercise of power and the resistance to which
it might give rise. Such arguments and struggles, though, are
seldom equal. They have . . . a political sociology that emerges
from their place in a system of relations. And so, as . . . some
people and practices emerge (or remain) dominant, their
authority expresses itself in the apparently established order
of things (Comaroff and Comaroff 1991, 18).

According to the Western definition, RUF and CDF children, boys and
girls, were all child soldiers: they were all exposed to the trauma of war.
Yet in postwar practice they are positioned and treated quite differently by their communities and by NGOs. Westerners have sought to
include all "war-affected" youth under the protective umbrella of their
interventions, but some distinctions resist that inclusion. In this chapter I show that RUF and CDF children have vastly different access to
the child soldier identity, and that boys and girls have vastly different
access to the child soldier identity. This state of affairs flies in the face of
the universalizing discourse of youth innocence and reveals in greater
detail the ways a globalizing model of childhood intersects with already
existing ethnic and gender dynamics.

The last chapter showed that there are real differences in the ways
formal and informal reintegrators can gain access to or make use of the
"child soldier" identity. That is, they can differentially not only access

benefits from NGOs but also discourses of abdicated responsibility and therefore forgiveness. In the same way that the informal reintegrator troubled the assumption of a unilinear reintegration trajectory that put child soldiers through a rehabilitation machine and back to normal life, the examples of CDF child soldiers and girl child soldiers trouble the iconic image of the abducted rebel boy. This chapter investigates the differences between RUF and CDF child soldiers, and male and female child soldiers.

Each of the two parts of this chapter should be considered within the same analytical framework: what happens when the child rights discourse bumps up against—in the case of the CDF—"tradition" (as a political system), and then—in the case of girl soldiers—traditional gender relations. I am interested in the intersection of different, at least partially incommensurate, models, and in determining what happens in practice at this particular historical and political juncture as those two systems come together.[1] These comparisons allow us to look critically at the power of child rights discourse as it is differently enacted in different structural locations, in relation to other preexisting discourses and power structures. This work is related to the work of anthropologist Sally Engle Merry and others who are investigating the ways human rights discourse is "vernacularized" across multiple contexts (Merry 2006). I am not merely pointing out variety; I am investigating the relative power of different constructs in practice, and investigating the changing contours of "youth" in postwar Sierra Leone through the lens of child soldiers.

RUF and CDF

The Civil Defense Forces (CDF) are groups of locally organized militias, emerging from and reconstituting hunting secret societies. Different ethnicities had their own branch of the CDF with the Kamajohs for the Mende, Gbethis for the Temne, Tamaboro for the Kuranko, Donsos for the Kono, and even hunting societies for the Krio and others in Freetown. They reference their traditional-ness and local-ness in explaining their power. The clothes they wear (at least for journalists) are *ronkos*, shirts made of native cloth, covered with charms and claimed to have the power to repel bullets and other supernatural powers (Richards

2009; Wlodarczyk 2009; Hoffman 2011). In truth, most of the time they dress like anyone else.

There is a growing literature on "hunters," and I am not the first to make the point that they are both traditional and modern, and, obviously, to complicate that distinction (see Muana 1997; Leach 2000; Ferme 2001a; Shaw 2003; Ellis 2003; Ferme and Hoffman 2004; Wlodarczyk 2009). Many scholars have written that the distinction between the "modern" RUF and the "traditional" CDF is specious, as the RUF drew on "magical" forces[2] and the CDF is a modern transnational force. For example, some of the arms used by CDF fighters were purchased by Sierra Leoneans in the international diaspora in a system organized over e-mail networks. I would rather characterize each faction as somehow fractal in its relationship to tradition and modernity. That is to say, behind each modern-seeming element, there is tradition; and behind each traditional-seeming element, there is modernity. In this way, I see the two factions as remarkably similar in cultural makeup. What becomes interesting, then, is not to determine which faction is traditional and which modern, but to see how various actors deploy the distinction between modern and traditional for various purposes. That is how we come to see the power of such distinctions. This is in line with what geographer and anthropologist Melissa Leach suggests as the way to understand the phenomenon of "hunters":

> If one is to understand the contemporary hunter phenomenon, (there is a) need for a theoretical lens attuned to discourses and representations about hunters—whether forwarded by hunters themselves or other players—and to their everyday practices and performances. (2004, vii)

Anthropologist Rosalind Shaw, for example, shows how that distinction is deployed in Western journalism, with results for U.S. foreign policy:

> In the print media's longer features on the ten-year war in Sierra Leone . . . what I call juju journalism became part of an established genre. Here, the mingling of "modern" technologies such as AK 47s with "magical" techniques has been taken at best as a sign of 'deep weirdness' and at worst as evidence that processes of counter-evolution are at work in the collapse of African states such as this one. (2003, 81)

Shaw concludes, "Juju journalism provided a convenient rationale for an isolationist politics of knowledge in relation to African conflicts more generally in the 1990s" (2003, 102).[3]

Scholars writing on "hunters" in the special issue of *Africa Today* edited by Melissa Leach (2004) have explored the deep and sometimes contradictory political position of hunters in their own national contexts. In my experience, these political differences come clearly to the fore in the postwar period with respect to the provision of demobilization, disarmament, and reintegration (DDR) benefits for ex-combatants. It is often claimed that former fighters of the CDF are not in need of DDR benefits, since they were never separated from their communities. The CDF responds angrily that the rebels were rewarded for their atrocities, whereas the CDF, who should be seen as national heroes, received nothing.

What scholars have not discussed, generally, is the role of children in the CDF. The CDF, when it is trying to be a modern political force, denies the use of child soldiers. But commanders on the ground explain their reasons for initiating children, and in some ways preferring children as members. The children who fought with the CDF saw frontline action, yet because they were generally perceived as fighting on the side of the government, and held up as heroes in their communities and in national discourse, they are believed to have less trouble reintegrating into their communities of origin. Therefore, less money has been made available for their education and other benefits promised to the young people who fought with the rebels. They were young, and they participated as combatants in the conflict, and yet, somehow, they are not "child soldiers." They are often judged as not needing the same interventions, including education, vocational training, therapy for post-traumatic stress, and so on.

Because the CDF are thought to be more "traditional" and the RUF more "modern," CDF child soldiers are differently located within the child soldier discourse and that of child rights in general. A child CDF member is a contradiction, since initiation into the society, in some ways, automatically makes him an adult. Therefore his status as child is doubtful. That is, the categories of "child" and "CDF member" are mutually exclusive within the logic of secret society initiation—though CDF commanders did not hesitate to take advantage of the UN's rules about disarming children, and presented children for benefits even without arms (see Ferme and Hoffman 2004, 88).

The sections that follow describe the recruitment, participation, and "reintegration" of CDF children, in my field sites with CDF presence.

Masakane

Masakane is a small Temne village in Masimera chiefdom, near the very middle of the country. It grew up on the main line between Masiaka and Mile 91 at the junction between the highway and the road to larger towns off the main road. This is a strategic location as any goods traveling from Freetown to the south or the east (Bo or Kenema) must travel along the fairly well paved road.

By 1993 or 1994, rebels started moving into this territory in part because they were being driven from the south by the activities of the nascent Kamajohs. They set up a base camp called Camp Fol Fol about fifty miles away. Because of its location at the base of a hill on the freeway, Masakane was the site of numerous rebel ambushes. As container trucks moved from Freetown with goods—everything from the staple food, rice, to expensive stereo sets—they would have to slow down around Masakane to gear up for the hill. The rebels took advantage of this and set up continuing ambushes there. As a truck would slow, rebels would jump from the bush, shoot out the tires, kill the driver and perhaps some of the passengers, and loot the goods. They would usually then burn the vehicle and abduct people from the area or people from the vehicle to carry the looted material on their heads the fifty miles to the base camp. This village became well known around the country as a very dangerous spot, earning the nickname "Foday Sankoh's garage" to describe the many burned-out shells of vehicles littering the ground there. The inhabitants told me they counted seventy-three burned-out vehicles along this stretch of road.

For a while the people stayed on, moving to nearby villages away from the main line. They thought the SLA would protect them and did not realize for some time that the SLA were involved in the attacks. According to my hosts, the rebels burned the village completely on April 9, 1995. Eventually, everyone fled the area and moved from camp to camp, most ending up in internally displaced persons (IDP) camps near Freetown. Pa Amidu, the headman of the village, and his family ended up at the Approved School Camp (named for the school grounds they took over) together with most of the people from that village. They then took part in creating a new camp at Grafton near Freetown to ease overcrowding at the first camp. The village, which had consisted of mud-block houses with occasional cement verandas and corrugated zinc roofs for the wealthy few, reverted to bush.

The men of the area organized to fight back. At first they hired Kamajohs to fight but had to pay them the equivalent of fifty dollars a day, more than they could afford for long. Some of them joined the Kamajohs and realized that the Kamajohs were using the same "leaf" in their ceremonies as the "leaf" they knew from their own society.[4] They realized they did not need to be part of the Mende-dominated Kamajohs and that they could form their own group. That is how the Gbethis—the Temne CDF—began. They became an important force in the area and most men and boys were recruited into the society. The head of the Gbethis has several wives and several households in the area, but Masakane served as a kind of base camp for him. The chance to interview him was another reason I picked Masakane for fieldwork.

The inhabitants of Masakane were in camps near Freetown for the 1997 junta period and also for the January 1999 invasion of Freetown. Their camps were overrun both of those times. By 2000, the men started returning to the village to see about rebuilding the houses and clearing the farmland after five years away. The history of this process is clear in the physical geography of the place. Walking around the village, one sees layers of construction and destruction. Alongside the burned-out shells of container vehicles, there are the burned-out shells of cement houses, sporting the frames of formerly grand verandas. There are the stick-and-wattle lean-tos that the men built first on their return. Farther back from the road are the traditional stick-and-mud houses with thatched roofs, now housing whole families of grandparents, women, and children. All of these layers of recent history exist side by side as a constant reminder (see figure 5.1, showing layers of destruction and construction in Masakane). There are now about twenty houses in the village and a small Roman Catholic primary school built out of sticks and palm fronds. There is one teacher for about one hundred children.

The women told me that when they finally returned to the village, they found strange skeletons in the fields, the remains of people who had tried to escape the rebels but had died in the bush. There are spent shell casings and other physical reminders in the ground around the village. My hosts used a metal army helmet, probably left behind by a Nigerian, to heat water over the fire. Other things are different in the village now. For example, the headman's wife, Fatmata, held an important position in the IDP camp, and some of that clout has transferred to their new life back in their old village. Also, Fatmata learned how to bake bread in an oil drum oven over a fire while she was at the camp. Now she bakes bread and sells it around the village or to people in passing vehicles.

At the time of my visit, in 2001, there was not really enough to eat. They were still waiting for their first harvest after five years away. The population of chickens and goats had yet to be replenished. "Bush yams" or indigenous roots were an important supplement to their diet.

The population of child soldiers in Masakane is almost all Gbethis, with a few stories of brief abductions by the RUF.[5] The boys were initiated into the Gbethis as part of the community. Girls were not involved. Some of the boys told me their role was to follow along after fighting and kill the wounded with machetes. In a way, the Gbethi child soldiers are caught in the middle. They were not demobilized as RUF and therefore do not get the same benefits as former RUF soldiers, but they also do not have the powerful political machine of the Kamajohs behind them. They may be the least visible of the child soldiers in Sierra Leone.

Only a few child Gbethis have gone through the formal demobilization process, but Caritas Makeni, a local child protection NGO, has registered some of the child ex-combatants and is providing some educational assistance.

CDF Children: Patterns of Recruitment and Participation

Recruitment into the CDF depended on existing local networks. Often, when the call went out for people in an area to join the CDF, all able-bodied men and boys turned up. Chiefs and other "big men" enlisted their own children. This is an interesting point because it contravenes the conventional wisdom in child protection circles that children separated from their families, street children for example, are most likely to be recruited into war. On the contrary, CDF children took part precisely because of their family connections.

In my interviews with former combatants, I found some variation in the pattern of participation between child members of the RUF and the CDF. In the RUF, child fighters were often on the front lines as a kind of human shield or first line of defense. In the CDF, child fighters often followed at the rear, their task to finish off the wounded enemy with machetes. The CDF sometimes used children as seers, claiming that some of the youngest and most innocent had more magical powers and could see through the enemies' supernatural protection.[6]

According to anthropologists Mariane Ferme and Danny Hoffman, Kamajohs'[7] identity was largely defined in the negative: we are the ones who do not do what soldiers do—namely, turn against the civilians, whom a military force is created to protect (2004, 80). The result was a lower incidence of abuses committed by the Kamajohs than by their counterparts in other factions, despite a similar demographic profile (though with the end of the war, more CDF abuses than originally suspected were uncovered, particularly in cases where Kamajoh units were deployed away from their home communities). CDF forces were also involved in various wartime atrocities: murder, rape, looting, and checkpoint shakedowns (though, admittedly to a lesser extent than the rebels). In fact, the leaders of the CDF were the first to come before the Special Court for Sierra Leone set up to try those most responsible for war crimes (see also Humphreys and Weinstein 2004a).

However, in general, the civilians saw the young combatants of the RUF and CDF as coming from the same segment of society. People told me, "They are all the same boys. They acted the same at checkpoints. They dressed the same and took the same drugs." In some cases, they were, in fact, *exactly* the same. I know of several examples of boys who had been through formal demobilization from the RUF or AFRC who immediately upon returning to their home village joined the local CDF, completely undoing their pledge not to pick up guns.[8] Exhausted civilians often told me *"sojaman no good—once he dae carry gun, he no good"* (Soldiers are no good. As long as they are carrying guns, they are no good). In practice, faction often mattered little in the everyday lives of Sierra Leoneans, and indeed any young man with a gun was a man to be feared.

In order to provide some ethnographic specificity, I present here three stories of CDF children drawn from my field sites: the Gbethi

Figure 5.1. Layers of destruction and construction in Masakane. Standing in the main road, in the foreground, is the rusting carcass of a looted truck, in the next plane is a temporary mud hut built by the first men returning to the village, in the next plane is the remains of a concrete house from before the war, and in the background are the more substantial mud and thatch houses in which people live now.

children in Masakane, the semi-formal reintegrators of Pujehun, and the boys of Brookfields Hotel in Freetown.

GBETHI BOYS IN MASAKANE

Before I ever met Obia, the Gbethi commander in Masakane, I had heard about him from some of the NGO workers. They told me he was a powerful man in terms of magical abilities, and that although he was only in his thirties he already had six wives. One of the NGO workers, an old friend of mine from my Peace Corps days, said he was excited to finally meet this Obia, someone so young and yet so powerful. And yes, Obia did have magnetism. Maybe it was the obvious deference of those around him, or that he was so soft-spoken. He kindly agreed to sit down to a lengthy taped interview with me.

Obia gave me the history of the war in the Masakane area, and the story of how the Gbethis grew out of the Kamajohs. The Gbethis are Temne and the Kamajoh are Mende, but the groups grow out of the same secret society tradition. "We use the same leaf," Obia explained. Obia is clearly caught between two systems of thought on the issue of child soldiers. He is a powerful man in his own context, but he also made sure to point out to me that he has a secondary school education and three O-levels, a high attainment in this region, the only chiefdom in the country, I was told, without a functioning secondary school. First, he explained to me all the reasons why it made sense to initiate children into the Gbethis:

- At the time they began, they thought the war might never end and they needed all the manpower they could get. (At the time, he explained, they did not know "the whites would come and save the war.")
- A Gbethi's strength comes from the length of time he has been a member of the society, so if they start initiating boys young, by the time they are older they will be at full strength, able to become invisible and so on.
- Young boys have an easier time keeping the laws (against sleeping with a woman not their wife, certain dietary restrictions, no drinking or drugs) because they are not used to those things.
- Some of the boys "*get yai*" (have magical powers, in particular the power to "see" the "devil" at the center of the society's power) and so are necessary to the cause.

On the other hand, the NGO people had started explaining to him why it is wrong to use child soldiers, and had brought aid for some of the children. So, he was able to explain to me that he knows the use of children in war is wrong, but that he did not know any better at the time.

It is also interesting to note that although he went along with the child protection people when they were around and registering child combatants, he continued to use several child bodyguards in his own retinue. Although I was told practically every boy over eight in the community was a member of the society, it was carefully negotiated between Obia and the child protection NGO who among the group would be registered as an official child ex-combatant. The child protection NGO could not afford to support all of the children who had technically been CDF, so they went into each village with a pre-set figure of how many they could register in that spot. They had their own system of ensuring that the boys had really taken part in fighting: Had they been on the front lines? Could they handle a gun?

Similar to the process for registering informal reintegrators for benefits discussed in the preceding chapter, the village and society elders had their own strategies for deciding whom to put forward for registration. Often it was the sons of big men in the village who got registered. (Obia's own son was a registered ex-combatant, though he still carried a gun for his body-guard duties.) The other thing taken into account in the calculation was the possibility of also registering as an adult for the more lucrative adult DDR benefits. Anyone old enough to possibly qualify for adult benefits would not register for the child program. However, for some reason the child protection NGO would not take children who were too young, so there was a delicate balance. Finally, the main sort of assistance the child protection NGO was able to offer at the time was UNICEF-sponsored aid to schools for each child ex-combatant (the CEIP, discussed earlier). There were plans in the pipeline for a skills-training center for those who were not interested in formal schooling, but that was in the foggy and distant future, and might never come to pass. Therefore, it was also ben-eficial to register as ex-combatants boys who were interested in school and, if possible, already attending the local school.

This means that the set of CDF boys formally registered as ex-com-batants bore some strange relation to the set of boys who actually were ex-combatants. Similar to the situation I described in chapter 4, usu-ally those who were most connected, already going to school, and most "integrated" were those who were rewarded with reintegration benefits.

The actual CDF boys I met described a range of experiences. Most said they had enjoyed being part of the action and were proud of their service to the community. One boy was very quiet, and said he was still disturbed by the images in his head of the killing he had been required to perform. I asked whether he had mentioned this to the child pro-tection worker, but he said he would be ashamed to. He was, after all, Obia's son.

THE CDF BOYS IN PUJEHUN

As I described in the preceding chapter, the boys of interest in Puje-hun—a Kamajoh-controlled area—started as RUF and then, upon their return after demobilization, were immediately recruited into the CDF.

Their recruitment into the CDF could be seen as a most powerful "reintegration" into their local community. The community's concern in their case was clearly not that they were child soldiers, but that they had been fighting for the wrong faction.

I learned a great deal about the Kamajohs as a political force while I was in Pujehun. The local commander was interested in showing me off in order to bolster his political clout, and he took me on long trips in his Jeep, explaining his take on the history of the war in that area. I also got his version (similar in many ways to Obia's) of why they used child soldiers at first but had now seen the light and had stopped—this despite the fact that I always saw groups of teenaged boys hanging around outside his office, sitting on the veranda ready to be "sent" at a moment's notice. Unlike Obia, who struck me as a "son of the soil," this fellow was urban and sophisticated. He talked about his family in the United States, and about plans for turning the Kamajohs into something like a parallel national army, the Territorial Defense Force.

The CDF boys I introduced earlier had used joining the CDF as a way of signaling that they were no longer affiliated with the RUF. The rest of the CDF boys were anxious to be associated with the most dynamic organization in the region. The CDF was clearly the site of political power in that chiefdom.

CDF BOYS IN FREETOWN

The CDF is supposed to be the rural militia, based on secret society ritual and magical power, but they occupied—and made their headquarters in Freetown, for over a year—the formerly posh Brookfields Hotel (see Hoffman (2005 and 2011, chapter 6) on his experiences at the Brookfields Hotel). Their larger political aspirations, to become a legitimate parallel army, made a foothold in Freetown essential. When I asked what they were doing in town, the reply was always "We're here to protect Freetown"—despite the presence of thousands of UN peacekeepers. Clearly they wanted to stay on the national agenda, and a presence in Freetown was the best was to do that.

I first met the CDF boys of Brookfields at a secondary school in Freetown where M. T., an old Peace Corps colleague, was teaching. He invited me to his school to meet the child soldiers studying there. I hung out at school with the boys, about fifteen of them, for a few days, and later they invited me to come around some day and see where they lived.

At the front gate there was a moment of confusion. "I'm here to see Ibrahim," I said. Instead I was taken to meet the commander and go through a ritual of deference. They took me into a small room, crowded with men. I sat down and explained calmly: I am just a student. I just want to talk to the boys. I do not have a gun. I support everything you are doing. The commander made clear that they were there to protect Freetown, and that they had not yet disarmed. He took a gun out of a desk drawer and put it on the desk (to scare me?). Luckily, by this point, I had seen enough armed men that it did not seem that unusual to me. I just kept talking in Krio, soothing his ego. He was a busy man, so he let me go on to meet with the boys.

Like almost everything remotely posh in Freetown, the Brookfields was destroyed at some point during the fighting. I expected that. But passing through the lobby I was haunted by the image of how it used to be. The Brookfields Hotel was the first place we had been taken for our Peace Corps orientation, twelve years earlier. Where there had been big black leather chairs, now there was nothing. There were graffiti on the walls about the power of the fighting force. Women were cooking on three stone fires in the parking lot outside the balconies. All the landscaping had been torn out. It looked, in fact, like a Sierra Leonean village, picked down to bare dirt—a clean yard, snake-free. There was no electricity here, as in most of the city, so the long corridors were eerie. As we passed through rooms I called out, "Oh! This used to be the dining room!" "Oh, there's the swimming pool" (no longer full of green water, now with a scattering of dirt and weeds at the bottom.)

A little room near the pool had been converted into a hangout for smoking *jamba* (marijuana). They call it "the ghetto." There were Rasta graffiti on the walls: "Rastafari," "Jah Love," "Jah Kingdom." The man who ran "the ghetto" was a former CDF combatant with only one arm (I assumed it was cut off by the rebels, but found out later that he had a gangrenous gun shot and had to have it cut off in a hospital). We talked for a long time. He was upset about his disarmament package, he was upset with the government and with his own commanders. He had been accepted into a skills-training program but could not afford the transportation to get to the skills-training center every day. And it seemed to me he enjoyed his life selling palm wine and *jamba*, and spouting a mix of political theories about the rotten system to the assembled boys. They had arranged for some special palm wine for my visit, so I sat and drank and met some of the other boys who did not go to school but still lived at the Brookfields. The war disrupted many an educational career. They explained to me that they had been used by the system and that they did not trust their own leaders anymore. The young men put it to me that they are the new generation, the conscious generation, taking an explicitly political posture.[9] Mostly I was surprised how much the political outlook of the CDF boys sounded exactly like the RUF and AFRC boys I had met in ICCs and in villages around the country.

My hosts next took me to their room. A room that had held two American Peace Corps trainees in 1987 now held eight high school boys. They slept in shifts. The furniture had been destroyed so they slept on mats on the floor, yet I think they thought of their surroundings as rather swank. They were very clear about wanting to stay in Freetown to further their educations. They realized their status as former child combatants was a kind of currency. Since funds from the NGO Children Associated with War (CAW) were drying up—the same problem discussed in detail in chapter 4 in relation to their programs in Pujehun—these young men were looking elsewhere for patronage. But at least they had made it to the city.

Back at school, and in the streets of Freetown, they wore the blue-and-white uniforms of their once-proud secondary school: a tradition, but sadly little else at that point. The school was falling apart. The teachers were rarely paid and make up the difference by charging the students for extra lessons after school. They bribed their way through exams. But they had the cachet of the blue-and-white uniforms. They would loudly boast and jostle at the food stands outside the gates of the school.

Conclusions Regarding the CDF Boys

The CDF means different things in different places. For the Gbethis in Masakane, being in the CDF was part of regular village life, for the boys in Pujehun it was an effective way to reintegrate after their time in the RUF, and for the Freetown boys being part of the CDF allowed them to try to claim "child soldier" status to get financial help with schooling. The lessons have to do with the self-representations of boys as they strategize about where they can get the best benefit package. The children themselves have to negotiate the ebb and flow of CDF politics and NGO funding.

There is a set of young men, facing the same set of problems, who through luck (good or bad) end up affiliated with one or the other of the RUF or the CDF. They may have joined the fighting for similar reasons and have done similar things during the war. They, in fact, see themselves as in the same boat. However, they have very different postwar trajectories. The CDF youth mainly rely on (already shaky) traditional ways forward in the face of a CDF leadership that, in its quest for modern legitimacy, denies their very existence. The RUF youth, on the other hand, mainly take a modern way forward, relying on the programs of international NGOs for training and relying on the Western ideology of youth innocence to make them new again. This is why strategic self-representation is key to reintegration trajectories.

A handful of youth confound these trajectories. For example, some RUF boys joined the CDF immediately on return to their villages, choosing the traditional route to reintegration. Some CDF boys in Freetown are desperate for education, and therefore do what they can to take on the modern child soldier identity and get NGO help. It is in their choices of strategic self-representation that they are working out the contours of the present-day distinction between modern and traditional, as well as the contours of childhood.

When CDF child soldiers make bids for educational support, it is as heroes—their service to the country means that they should be rewarded with benefits. In that sense, they do not make use of the discourses of abdicated responsibility that are the hallmark of RUF child soldiers. They rarely say they are traumatized (even if they are—see the case of Obia's son). This raises the larger question: how is it different to fight for the "good guys" than to fight for the "bad guys"?[10]

In a way, the CDF stands politically for the old-fashioned patrimonialism, and children and young men have their usual place at the bottom of that hierarchy. As such, they should not ask for anything, their needs are supposed to be met by the elders. This is contrary to the RUF political ideology of being anti-tradition, or at least anti-the-current-rulers.

In that sense, the RUF are "modern"—not because of their weaponry, not because they do not use magic (they do), not because they are the "loose molecules" of Kaplan's (1994) analysis (the CDF youth are the same "loose molecules"), not because they target civilians, but because of their political ideology against gerontocracy.

The CDF boys are in some ways in a worse position than the RUF boys. Look at the Gbethi boys in Masakane. They are stuck in the village with no rights to speak of, and all benefits accruing from their positions as former child soldiers exist within a political system in which they are on the bottom. The Brookfields boys show that there is dissatisfaction with this position, and with what has been promised them by their own elders. It is clear that they desire education and skills training and will do what they can to gain access to them. It is not just children who are dissatisfied. There is bitterness within the adult rank and file of the CDF as well. With time, as dissatisfaction grows, there may be less difference between former RUF and former CDF as they unite against the slow provision of government benefits for ex-combatants.[11]

The real lesson has to do with the central argument of this book. It is to be found by tracking the self-representations of boys as they strategize about where they can get the best benefit package, whether within traditional village structures, or in the new modern CDF, or by casting themselves as "child soldiers" in order to get educational support. It is the interplay of structures and strategies that in part determines the shape of the identity child soldier, and hence the progress of modern childhood.

Boys and Girls

Anthropologist Chris Coulter's 2009 book *Bush Wives and Girl Soldiers: Women's Lives Through Peace and War in Sierra Leone* is built on methodological and theoretical choices similar to my own. Coulter spent a total of fifteen months in northern Sierra Leone, spending time with women who had been abducted into the rebel forces. (In the Sierra Leone case, girls and women were associated mostly with the RUF and some SLA, but in general not the CDF.)[12] Based on her informants' stories, she provides details of their abductions, their participation, and their struggles to fit in to postwar life. As I found for child soldiers generally, she discovered that there are different patterns of association: some were abducted, some joined for protection, some seem to have joined at their parents' urging. There are different patterns of experience: many were raped, and used for sex, though some report forming close relationships with particular commanders who protected them from other combatants. Some were domestics, cooking and cleaning in the way they might in normal life. Some became fighters, and sometimes even respected commanders (Bah 1997; Mansaray 2000; National Forum for Human Rights 2001; Coulter 2009). Some of the boy soldiers I interviewed told me that some young women who were the "wives" of commanders had great power around encampments and hardly had to work since they had legions of young boys to work for them.

The story of Aminata, the daughter of Pa Kamara in Rogbom,[13] reveals the complex nature of participation and some of the motivations for involvement with the fighting forces. Aminata was raped in early 1999. At that time, she was about fourteen years old. Her father decided that it would be in her best interest to give her to a woman collaborator, Mammy Haja, who traveled with the rebels, selling them drugs and supplies they needed but could not loot (tinned milk, MSG, cigarettes, and so on.) Aminata told me the story this way:

> She said, "Pa, I want your child. Why don't you give her to me now before another man goes and holds her." So, then the Pa said, okay, I will tell her mother. . . . Because at that time, they had already held *all* my fellow girls. Men held them all and carried them away to Blama. So my mother said okay, before rebels come hold me and carry me away, let him leave me with the woman. . . . He left me in the woman's hands. But at that time, when I was with the woman, I wasn't with any man again. I was in the woman's hands.

Aminata's father told me that he assumed the separation would not be for long. As the rebels moved on, Mammy Haja went along and took Aminata with her. They also took all of the young boys and young men to train. The adults of the village were heartbroken, not knowing when they would see their children again. During this time, Aminata was often sent into Freetown to buy drugs—mostly a crude form of heroin ("brown brown") and some cocaine—to bring back to the fighters.

There were checkpoints, but I didn't have any problem passing. If I came down inside a checkpoint where they checked they would say, "Small girl, pass." I passed. They didn't used to check me. So they didn't know that I was coming from the rebels' place.

Aminata knew if she tried to escape she would be found out and possibly even killed. Eventually, her family was able to secure her release, and now she is attending secondary school with her fellow ex-combatant children. She says those days are hard to forget. She still sees some of her former captors, men who have been retrained as members of the national army, manning check points near her village. She never went through a formal demobilization program, so she is not eligible for the benefits that NGOs offer other child ex-combatants.

Aminata's story is interesting for many reasons. First, there are the issues of abduction and sexual abuse that are familiar in stories of female child soldiers. But in this case, she was not exactly abducted. Rather, her father turned her over in an effort to protect her. She never carried a gun, but played a supporting role. She was not with a rebel commander, but moved along with the rebels, with someone who saw herself as an adoptive mother. Hers is not a clear case of victimization, although clearly that plays a role. Aminata and her family demonstrate personal agency and make strategic decisions during and after the war. Her main concern now is how to get registered somehow to get some of the benefits she sees others getting.

Girl Soldiers as a Problem

Over the past ten years, interest in girl soldiers has risen among NGOs and some academics, with the understanding that girls are perhaps the most marginalized of all ex-combatants; among others, see West (2000); Mazurana and McKay (2001); Shepler (2002); Mazurana et al. (2002); Keairns (2002); McKay and Mazurana (2004); McKay et al. (2004); Utas (2005a); Schroven (2005); Park (2006); and Coulter (2009. Susan McKay rightly notes that the issue of girl soldiers is sometimes seen as primarily a child protection issue and that at other times girls are inappropriately grouped under the larger rubric of women, but their particular situation as *girls* is generally not central within either child rights or women's rights discourses. Child protection researcher Gillian Mann agrees that studies into the situation of separated children may miss the experience of girls after they reach the age of twelve or thirteen, since after this point their needs and experiences may be understood as those of women and not of children (2004, 17). Susan McKay suggests that "girl soldiers not be categorized as the province of any particular group because the potential exists that others will abdicate responsibility to act" (McKay 2006, 91), and that instead we need to understand girlhood as its own category.

Girls are now often mentioned as a priority in UN documents, starting with UN Security Council Resolution 1325 on Women, Peace, and Security. The Capetown Principles state that "particular attention should be paid to the special needs of girls and special responses should be developed to this end" (Legrand 1997). The more recent Paris Principles (UNICEF 2007) contain a separate section on "the specific situation of girls" (Section 4). Despite the recent focus, however, there are still relatively few programs for girl soldiers and little is known about their experiences during and after war.

In Sierra Leone, some girls went through the formal system, showing up in interim care centers, but they were few in number, and they were not well served.[14] Some girls ended up with their demobilized commanders in DDR camps as "camp followers" rather than demobilizing on their own. UNICEF staffer Andrew Brooks reports that girls were kept under particular control and represented a presence among "camp followers" that contrasted sharply with their absence in interim care centers (2005). In ICCs, they always stood out among the legions of rowdy boys. I met several ex-combatant girls in the gara tie dyeing class I took at Lakka. We were all rank beginners and spent long hours learning to tie "razor blade" and "cow foot" patterns. Mariama had trouble concentrating on the gara patterns and insisted that she was a "civilian" despite common knowledge that she had been "married" to a high-ranking RUF commander; Fatu was aggressive and manly in her clothing and carriage; Rugiatu was serious about her school work and about finding her family near Makeni; Grace bragged to the boys—to the boys' amused disbelief—that she had been involved in the 1999 invasion of Freetown.

In Sierra Leone, although many estimate that roughly equal numbers of girls and boys were abducted by the rebels, the percentage of girls in formal demobilization programs was about 5 percent of the total, hence the difficulty in collecting reliable data on the population of girl soldiers. Most of what is known about girls associated with the fighting forces who did not go through a formal demobilization—and the majority of them did not—is through interviews with girls who joined skills-training programs well after formal DDR was over.[15]

Girls Have Different Postwar Trajectories

In Sierra Leone, girls' reintegration is different from boys', and it is structured by gender. As Chris Coulter explains, "Many were afraid to return home, fearing rejection by their families and communities. With good reason, they were afraid of being punished for returning with rebel children, for not being virgins, and for being called rebels" (2009, 241). Girls face an explicitly moral discourse about their participation in the war. The key point is that girls (and women) affiliated with the fighting forces in Sierra Leone have different possible trajectories of identity to follow in their postwar remaking.

In their struggle to be accepted back into their communities, boys use discourses of abdicated responsibility: "I was on drugs," "I was abducted," "It wasn't my choice." However, I rarely heard girls making use of the same discourses of abdicated responsibility. Although, in practice, girls often had very similar situations—they were abducted just as the boys were—there is some degree to which sexual activity, even rape, is perceived to be their own fault, or at least something which cannot easily be undone. As Binta Mansaray notes, "For some women, life will never be the same; while men can move on, remarry and start new families, women victims of rape have no such chance. Although they are victims, their lives are forever marred by the social stigma associated with rape" (2000, 143).

This means girls use different strategies for their reintegration. It means they are more likely to slink home, perhaps pregnant or with a baby, and try to keep the whole thing quiet. Their strategy is secrecy and hope for eventual marriage. Anthropologist Caroline Bledsoe, writing about Mende girls' marriage strategies, discusses women and marriage as currency for patron/client hierarchies. She explains, "Since young women bear valued children and provide most subsistence and household labor, giving them in marriage has long comprised the cornerstone of families' efforts to create obligations toward both potential patrons and clients" (1990c, 292). Families may collaborate in secrecy in order to keep a girl marriageable. Chris Coulter provides a good description of how marriage practices are changing in Sierra Leone, especially after the conflict (2009, 74–94), but the fact remains that marriage is an important social institution involving families in networks of reciprocity, and a key part of girls' and women's own notion of what a "normal" postwar life entails.

Another reason that girls may not enter formal reintegration programs is that the primary long-term benefit offered by formal reintegration programs is help with schooling. Most girls in Sierra Leone do not attend school, especially those who may have missed years of potential schooling while "in the bush." Formal programs simply do not offer the sort of help girls feel they need to achieve what they want: respectability and marriage potential.

There are skills-training programs in soap making and gara dyeing, but there are many critiques of such programs: the explosion of similar skills training programs around the country means that the number of people with those skills will easily exceed demand, production of these products requires an investment in raw materials which may be difficult (Coulter 2009, 192). Of course, similar problems have been noted with skills-training programs for boys, but again we come up against the notion of different social trajectories for boys and girls as a result of the traditional sexual division of labor in Sierra Leone. It is easier to imagine an ex-combatant boy, trained for example as a tailor and possibly provided with a sewing machine, making a new identity as a local tailor. It is rare to see a woman who can support herself solely through such individual industry. The exception is what Coulter calls "loving business" (2009, 199) and anthropologist Mats Utas calls "girlfriending" (2005c), clearly a gendered livelihood strategy with associated moral valuation.

Young mothers are especially vulnerable; their babies are sometimes perceived to be the "rebels" of tomorrow.[16] Some are rejected by their communities, while others leave ashamed of their failure to fulfill the roles expected of them. Interestingly, some agencies are addressing the problem of child mothers by, in some cases, encouraging girls to marry their former commanders and captors.[17] People generally realize this is a delicate position, but the point is that, culturally, marriage somehow solves the problem of reintegration for girls in a way unavailable to boys. No one would suggest that boys formalize their relationship to their erstwhile captors.

There is also discussion among some NGO staff of trying to get former rebel boys and girls to marry each other. The staff of the Lakka ICC were very proud of the fact that two of their hardest cases—a boy and a girl, formerly fighters with the RUF—had been married by the priest who ran the program and were trying to make it on their own. I heard this strategy echoed by ordinary members of the public as well. In a debate with a young Sierra Leonean lawyer I knew, I asked whether it seemed fair to him that "rebel girls" were less easily accepted into society than "rebel boys." He agreed that it was not fair, but that it was unavoidable. "I wouldn't want one of those girls marrying into my family," he explained. He suggested the solution that the rebel boys and girls should be married to each other, and then there would be no stigma (see also Coulter 2009, 227).

There is also some positive change for these girls post-conflict. In particular, some teen mothers do return and are accepted by their families and communities. I met the mother of a teenage girl abducted by rebels who told me proudly that she had accepted her girl back, that she was raising the baby as her own, and that after several years' disruption in her schooling, her daughter was now at the top of her class. This population of girls is changing some of the old rules that say that once a girl gives birth, she can never attend school again.[18] Although I have talked a lot about the unequal impact of a moral discourse, and the traditional expectation that girls should just get married, one of the unexpected results of the war is that the practices and ideologies surrounding youth and gender in Sierra Leone are changing to some degree in response to this population of war-affected girls. Certainly feminist scholars have pointed to the postwar moment as a time when gender roles are thrown into question, and the possibility exists for genuine social change (for example, El-Bushra 2003).

Gender Inequality as Structural Violence

Although there have been some small changes, the power relations inherent in the patriarchal and gerontocratic system in Sierra Leone were not produced by the war, and to a large degree they exist unaltered after the war. It is imperative to draw the connection between gender inequality as structural violence and the more spectacular exceptional violence of war (Cockburn 2004). Others have noted continuities between the atrocities of the prewar era and the atrocities of the civil war era (Richards 1996; Ferme 1998), but it is also possible to draw that connection forward into the postwar period.

The first example is the well-publicized February 2002 study by UNHCR and Save the Children–UK in which displaced and refugee children told investigators that aid workers and some security forces extracted sexual favors in exchange for food and other services (Save the Children–UK 2002). This was widely reported in the Western media as the sex-for-food scandal. There was shock in the Western media, and Kofi Annan demanded follow-up actions, but no Sierra Leoneans I know were surprised to hear of it.

The second example is the widespread phenomenon of girls and women attaching themselves to ECOMOG and UNAMSIL troops, sometimes at the urging of their parents. It was reported in the national Child Protection Committee meetings organized by the Ministry of Social Welfare, Gender, and Children's Affairs, that there was a problem with young girls hanging around outside UNAMSIL camps up-country in hopes of connecting with some rich UN peacekeeper, or at least of trading local produce for UN food supplies. The UNAMSIL representative at the meeting denied this was happening, but again the Sierra Leoneans did not seem surprised. "What can we do?" one asked. "Most times it is the girl's parents who have sent her there."

A third example: I was on a big public transportation truck, traveling to Freetown from up-country. We stopped at one of the many checkpoints along the way for the driver, through his apprentice, to give the police "a little something." Two ECOMOG soldiers carrying guns approached the driver for a lift for a fellow Nigerian soldier and his girlfriend. What could the driver do but comply, and the two new passengers hung on near the doorway. The girl could not have been more than thirteen, and she was dressed skimpily and giggling. The soldier shouted and waved to his colleagues along the way. My fellow passengers started commenting on this sorry state of affairs in a way that the girl could understand but the man could not. She just smiled back at them. When the two came down from the vehicle, the passengers erupted in critique: "What is wrong with our youth today? Why do they live such useless lives? Why don't they value education?" However, there was no criticism of the ECOMOG soldier.

There is a continuity here that must not be ignored. The problems of girls in Sierra Leone did not start and stop with the war, and understanding their "reintegration" requires understanding the situation of young women in Sierra Leone. What does it mean to be "reintegrated" into a system of such inequity? (MacKenzie 2009b). As I asked about informal reintegrators in general in chapter 4, how does one define a successful reintegration: Is it a return to the status quo? Or is there room for another model?

Humanitarians are missing a great deal by applying a normative framework, seeing these girls always as passive victims, and by not seeing the range of possible desirable outcomes that the girls themselves see. As Psychologist Erica Burman concludes about Western conceptions of children affected by war, "If the price of innocence is passivity, then the cost of resourcefully dealing with conditions of distress and deprivation is to be pathologized" (1994, 244). We gain a lot by close attention to how the girls themselves are maneuvering through the system and how girls themselves understand their ambivalent agency—what choices, if any, they see for themselves.

Conclusion

In postwar Sierra Leone, new definitions of youth are being forged in contradictory and extremely political ways. The main message of this chapter is that the global model of childhood—as expressed in the "child soldier"—only goes so far. For a variety of reasons, it "works" for some children (boys of the RUF) and not for others (CDF members, girls).

Investigating the differences between CDF and RUF children reveals quite a bit about the *re*construction of youth in Sierra Leone. Their case shows a contradiction, or fault line, along which youth is worked out. The children who fought with the Kamajohs and Gbethis saw front-line action, yet because they were generally perceived as fighting on the side of the government, and held up as heroes in their communities and in national discourse, they are understood to have less trouble reintegrating into their communities of origin. Therefore, less effort has been made and less money has been made available for their education and other benefits promised to the young people who fought with the rebels. They were young, and they participated as combatants in the conflict, and yet they are not judged as needing the same benefits, including education, vocational training, therapy for post-traumatic stress, and so on. This state of affairs suggests that the programs for child soldiers are not in place simply because the recipients are children who participated in war and are therefore particularly traumatized, but because they are "rebels" and a place is needed to hold them until something can be done with them. Are children somehow less traumatized when they fight for the "good guys"? Sierra Leoneans apply the discourse on the rehabilitation of child soldiers when it is politically useful—to help forgive the rebels—and do not apply it when it is not useful—to continue to exalt the civil defense forces.

Girls affiliated with the fighting forces are even less able to access the benefits, discursive and material, which come with the identity "child soldier." The postwar models of youth are profoundly shaped by the prewar models. The larger question is: Which globally circulating discourses are taken up by the society, when, and why? In particular, why is child rights discourse so popular in Sierra Leone after the war? I argue that it is because child rights discourse is valuable strategically, for the society's postwar rebuilding, but that value does not seem to extend to the possibility of changing situations for girls.

In Sierra Leone, gender constructs and the constructs "traditional" and "modern" are quite powerful, politically and for strategic self-representation. To reiterate, on one hand, according to global (modern) definitions of childhood and institutions for child soldiers, CDF boys and abducted girls are best understood as "child soldiers"—traumatized and used. But in Sierra Leone, neither group has been able to successfully claim that identity, either for easing their own reintegration, healing their own trauma, or accessing a set of benefits from the international community. We can only examine these countervailing cultural forces—modern childhood on one hand, "tradition" and gender as powerful social structures on the other—as they are worked out in specific localities at specific junctures by specific actors. An investigation of the strategic self-representation of former child soldiers, and asking which models of childhood gain traction in which situations, illuminates the processes of the globalization of modern childhood.

Conclusion

> Conceivably, the interventions I described in this book could
> be aligned into a narrative about improvement schemes
> becoming more effective, more people friendly, and more
> participatory. . . . Although I understand the temptation of a
> narrative that traces the improvement of improvement, I am
> not convinced by it (Li 2007, 274–275).

This book has been about children deployed in battle, but more centrally it has been about the deployment of the global ideals and modern techniques of childhood. Throughout, I have focused on differentiation in the population of child soldiers in Sierra Leone. One reason for this was simply to describe the complexity in a field where the conventional wisdom frequently universalizes the experiences of all child soldiers. But a more important reason was to begin to demonstrate some of the political effects of those differentiations and make clear that the techniques behind the creation of "child soldier" as a postwar identity have serious and unexpected effects. I began the book by describing the Sierra Leonean model of childhood, and its continuities and discontinuities with the participation of children in war. I then turned to a description of the rights-based child protection system that was deployed during and immediately after the war to deal with the population of traumatized former child soldiers, noting that people navigating the institutions reworked them, or vernacularized them (Merry 2006) to meet their own needs. In their practice they made real the very distinctions on which the system was based, and hence helped localize a global form: modern childhood. Naturally, this did not occur uniformly across the cultural landscape, so I have discussed in detail the differences for formal and informal reintegrators and for former combatants from different fighting factions and of different genders. These distinctions are not just inefficiencies in the application of programs, but properly seen can reveal the underlying politics in the spread of modern childhood.

In this book I have shown that in some ways Western interventions designed to ease the reintegration of former child soldiers in fact make that reintegration more difficult. To many Sierra Leoneans, what is needed is for child participants in violence to become mute and return to their place at the bottom of the social hierarchy, rather than to make new claims on resources. However, some young people reject this notion and refuse to go back to the dead-end social locations they inhabited before the war.[1] Some NGO practices harden the child soldier identity through labeling and list making, and provoke community anger at the inequitable distribution of benefits to child ex-combatants to the exclusion of other war-affected youth. Finally, the Western model depoliticizes youth, allowing a change from a previous model in which youth, and the potential of youth revolt at inequity in the patrimonial system, served as a check on abuses. Former child soldiers gain something—ease of reintegration and forgiveness—but they lose something as well—namely a kind of political agency that is denied Western youth. NGO activities purporting to help former child soldiers are in some ways buttressing patrimonialism and rendering obsolete previously existing forms of youth power.

Beyond sensationalist descriptions of the horrors of child soldiering, beyond even the rights-based approach to child protection, this book has been about social *practice*. First, practice as a way of resolving the divide between structure and agency that dominates so many anthropological studies of child soldiers. Second, and more importantly, practice as a way of understanding how global discourses (like "the rights of the child" or even "child soldier") function in the lives of their supposed targets. Scholar-practitioner Nicole Behnam writes eloquently about how teachers in Sierra Leone responded to "human rights flying all around" (2011, 101), claiming, "After the war in Sierra Leone human rights language was suddenly everywhere, and the insertion of these concepts into . . . society was unguided and frenetic" (89). Practice lets us see how some discourses "flying around" take root when others do not, depending, in part, on where they land. This is more than just an understanding of norm diffusion. Ethnography allows us to see global processes from the bottom up, where target populations strategize in relation to a swarm of new ideas and resources.

*   *   *

When I first started this research, in the mid-1990s, there was very little scholarly work on the subject of child soldiers. In the past fifteen years, the field has become more crowded.[2] I said from the beginning that my goal was not to undertake a policy review or a program evaluation, and indeed I have argued that the NGO perspective can blind us to the real lives of former child soldiers. But in this last chapter I turn briefly to the questions of policy makers and child protection programmers. There are some clear policy implications that grow out of my research. The most important of these is that policy makers must be cognizant of the political consequences of their distinction making. Then, and most obviously, it must be acknowledged that the simplistic use of one standard—what Rosen (2005) calls "the straight eighteen policy"—to determine who is a child and who is an adult is not really applicable in the Sierra Leone context. More complex models should be developed that better capture the social complexity behind the marker "youth."

Additionally, it is vital to work within already existing local institutions and frameworks. This means supporting already existing apprenticeship models rather than creating new school-like skills-training programs, supporting already existing (and struggling) community schools instead of creating new ones for specific target populations, and supporting fosterage of separated children rather than institutionalization in an ICC or other alternative care situation.

Finally, it is better to design programs to benefit all war-affected youth rather than to single out former child soldiers. As I have shown throughout, singling out former child combatants has serious unintended consequences. This orientation allows former child soldiers (especially girls) to use a strategy of secrecy for their reintegration, and it also acknowledges that structural violence can be as damaging to children as the more extraordinary violence of war.

I am happy to say that since my original research over a decade ago, the general trend in the young field of protection of children affected by armed conflict is moving in exactly these directions. Scholar-prac-titioners Lindsay Stark, Neil Boothby, and Alistair Ager at the program on Forced Migration and Health at Columbia University provide an excellent summary of trends in programming for the reintegration of children associated with armed forces and armed groups, including a discussion of the state of the research, describing what we know as well as the remaining knowledge gaps in "Children and Fighting Forces: 10 Years on from Cape Town" (Stark, Boothby, and Ager 2009). There have been significant strides over the last decades.

One of the best examples I have seen of culturally sensitive program-ming in action was a conversation between Obia, the Gbethi com-mander introduced earlier, and a child protection worker from Caritas Makeni. When they met to discuss the best interests of children, it was within a framework of two equally respected men, speaking the same language, drawing on shared ideas of what is best for children. It is not necessary (and in fact can be counter productive, as I have shown) to impose a Western construction of the child in order for Sierra Leoneans to care for and about their children.

Implementing the 2007 Paris Principles, for example, requires a deep knowledge of local contexts, and it requires allowing people to do the important work of postwar reconstruction using the "child protection" framework already existing in their culture. I am particularly pleased that international psychiatrists in particular seem to be coming around to this view. Psychiatrist Lynne Jones, in a summary article, recom-mends "greater attention to the child's perspective, their individuality and the cultural, social and political context in which they live" (2008, 291). She even recommends ethnographic methods as part of emer-gency needs assessment (294). As child protection researcher Gillian Mann contends, "By imposing systems of support that appear to out-siders to fit the local context, but which in reality may not recognize the specific content of existing practices, agencies can undermine tradi-tional support mechanisms for children" (2004, 19).

Ethnography and the Politics of Evidence

Child protection programs for war-affected children have historically been rights-based, starting with a list of universal rights and assumed needs of children. Standard measures and child welfare indicators have been important to this project of UN-sponsored measurements of childhoods against a Western ideal (Boyle 2010). In this young field, as in many other areas of development work, there is a new call for rigorous approaches to social policy and towards "evidence based" programming. The editor of the psychosocial forum of the Coalition to Stop the Use of Child Soldiers asserted in 2007,

> The question of the effectiveness of our interventions in reducing child distress is not primarily a theoretical question, but largely an empirical one. Stringent evaluations of interventions, including both positive and negative child reactions, can over time, result in the accumulation of evidence as to their effectiveness. Once the need for such an evidence base is accepted, there arises immediately a series of further questions such as: which approach works (is effective) for whom? What is the rationale for any given approach and how is its effectiveness to be measured? What kind of evidence are acceptable—for example, is it sufficient for participants to say they feel they have benefitted, or do we need additional empirically based outcome measures? . . . Clearly the accumulation of an evidence base for the effectiveness of psychosocial interventions of whatever kind must be a priority for the future (Dowdney 2007).

This begs the question of what kind of evidence is acceptable.[3] I argue in this concluding chapter that we can gain "evidence-based" and policy-relevant knowledge from ethnography that we cannot gain from other research methods. Of course, the use of ethnography in policy development is generally limited, and particularly so in conflict and post-conflict zones where issues include the safety of the researcher and the ethics of taking time for ethnography when there are pressing survival needs at hand. But ethnography is important because it can do things that other approaches cannot do.

First, ethnography can reveal the life worlds of children from their own perspectives and illuminate alternate indicators of well-being. Sociologists Alison James and Alan Prout, pioneers in childhood studies, include ethnography among the key features of the paradigm, stating that, "[e]thnography is a particularly useful methodology for the study of childhood. It allows children a more direct voice and participation in the production of sociological data than is usually possible through experimental or survey styles of research" (James and Prout 1997, 8).

Second, ethnography allows us an empirical basis from which to challenge the Western model of childhood. As anthropologist David Rosen puts it, "Ethnography—particularly the methods of participant observation—has unsettled conventional concepts of childhood and remains the best way to study children. Observing and listening to the voice of the child in natural settings, where children are not disempowered by the regimes of formal interviewing, testing, and measurement, provide the clearest portraits of the competence of children" (Rosen 2005, 133).

And finally, I still deeply believe that ethnography can contribute to creating more culturally appropriate and therefore more effective programs for war-affected children and youth.

Yet despite what I and others see as the clear contribution of ethnography to understanding child protection issues, the more valued method of research—among scholars as well as policy makers—continues to be quantitative, especially data from large surveys.[4] Although all kinds of research methodology can be useful, I interrogate here the preeminence of quantitative approaches (Andreas and Greenhill 2010), illustrating the weakness of relying too much on one kind of research with some specific examples from the world of programming for reintegration of former child soldiers. I compare some recent "evidence-based" work to my own long-term ethnographic engagement with conflict-affected children and youth to point to some areas where policy could be enriched by including the ethnographic perspective.

In some ways the embrace of ethnographic methods grows out of my frustration with some recent work on the reintegration of former child soldiers in Sierra Leone. Some public health researchers (for example, Betancourt, Pochan, and de la Soudiere 2005; Betancourt, Borisova et al. 2010) in particular claim to measure psychosocial adjustment using standardized measures of mental health outcomes administered through surveys. While acknowledging that "[i]n non-Western contexts, such as Sierra Leone, the cultural validity of constructs measured remains a perennial challenge for cross-cultural research" (Betancourt, Borisova et al. 2010, 1083), they nevertheless ask questions on protocols about terms that they assume have obvious meanings. It is important to understand how terms like "community acceptance" and "psycho-social adjustment" might have different meanings in different contexts, depending on whom one is addressing. For example, as we have seen, one of the findings of my work with former child soldiers is that they tell different stories to different audiences at different times. I came to see their behavior as strategic self-presentation, an important part of the creation and practice of postwar social identities. Therefore, in response to an official survey, a wide range of official answers may seem most appropriate to the respondent. In particular, I have shown that in postwar Sierra Leone, one might strategically exaggerate one's "trauma" to someone suspected of working for an NGO with the hope of some eventual benefit.

Another study concludes that stigma (manifested in discrimination as well as lower levels of community and family acceptance) is asso-ciated with psychosocial adjustment (Betancourt, Agnew-Blais et al. 2010). The researchers find that stigma is an important variable, and they study it as one of many possible symptoms. However, stigma does not exist in the heads of individual children or young people; it is a social and political phenomenon. Shifting understandings of the war over time will shift how community members relate to the former com-batants in their midst. As anthropologist Harry West, in his study of the long-term effects of girls' participation in war in Angola, argues, "I wish to call attention to the fact that social memory is sustained in the medium of ever-changing social contexts. Where trauma is held at bay in the moment of the experience of violence by the force of narrative accounts that frame the violence as purposive and meaningful, these narratives may shift and disintegrate when present-day realities under-mine their ideological claims" (West 2000, 182).

Finally, another standard component of vulnerability is commu-nity acceptance, often found to be a critical component of successful reintegration. But responses to a survey might be highly influenced by something that is happening in the community on the day the survey is administered (say, for example, the violence around the ICC in Lakka described earlier), but that context could not be taken into account in the research design.

There is certainly a need for longitudinal research on war-affected youth. For example, psychologists and child rights advocates Neil

Boothby, J. Crawford and J. Halperin (2006) carried out research on thirty-nine former child soldiers in Mozambique over a sixteen-year period. But too often long-term research means following individuals' responses to surveys over time. Especially in the field of psychosocial programming, the social component means that we need to include the social context. Indeed, postwar healing happens both in individual heads and in communities. Studying changing outcomes for particular individuals removes them from their context and, most importantly, depoliticizes both the conflict and the humanitarian response to it.

The "Child Soldier" Is Political

The problems of child and youth social reintegration are clearly complex and clearly political, yet policy makers seem to want to turn away from that complexity. Why is it that a randomized survey is more valuable to policy makers than ethnographic work? Why are policy makers so loathe to utilize ethnographic knowledge? I must say, now that I have worked in Washington, DC, for some years and have interacted with policy makers in various contexts, I have a slightly more nuanced sense of who they are and the constraints under which they operate. They often say they appreciate the anthropological perspective, but they perhaps do not know how to use ethnographic knowledge in policy and programming work. Ethnographic knowledge is not immediately "policy relevant" or "rendered technical" (Li 2007).

So we return to the question of power, and of what counts as evidence. I do not claim that one is better or worse, but I encourage thinking about why one type of data is more valued than the other, and this requires reflection on the politics of knowledge creation and of expertise. An ethnographic approach to understanding children's actual lived experience can contribute to more effective policy and programming that will help to support the "best interests of the child." Children's experiences should be included in efforts to understand social change—in this case, postwar rebuilding—and child protection policy should proceed from an understanding of children's lives in context rather than from a set of supposedly universal rights. This book has taken on human rights in its framing not because I oppose the goals of the human rights movement, but because I see it as unavoidably laden with power.[5] The best outcomes for children rely on us seeing Sierra Leonean childhood with our eyes open, rather than merely deploying the modern ideal of childhood.

NOTES

NOTES TO THE INTRODUCTION

1. They were all in the group of about five hundred children handed over in Makeni in May 2000. They told me they had been transported by the rebels from Kono (in the east) to Makeni (in the north) for the handover, probably in a rebel bid to downplay their presence in the diamond areas.

2. The actual number of children within the ranks of the fighting forces in Sierra Leone is impossible to calculate. For planning purposes, based on approximate numbers submitted by the factions, the National Committee for Disarmament, Demobilization and Reintegration (NCDDR) estimated there would be 45,000 combatants to disarm. Of these 12 percent, or 5,400, were forecast to be children. Few now dispute that this percentage is a gross underestimate (Brooks 2005). UNICEF Sierra Leone later came up with the estimate of 7,000, and the Coalition to Stop the Use of Child Soldiers estimates 10,000.

3. There is, of course, a countervailing common understanding of childhood in the West that sees children as increasingly dangerous. See, for example, the ever-younger ages at which children in the United States are tried as adults. See also the pioneering work of the Centre for Contemporary Cultural Studies on childhood and "moral panics."

4. The United Nations Convention on the Rights of the Child was adopted by the UN General Assembly in 1989. It was signed and ratified more quickly and by more states than any other UN convention.

5. That is almost a folk definition of hegemony: hegemonic forms are forms of power that seem like common sense.

6. This reminds me of the point that "alcoholic" is a term of identity only for those who no longer drink.

7. Andreas Reckwitz performs a useful genealogy of social practice theory in his article "Toward a Theory of Social Practices: A Development in Culturalist Theorizing." He explains, "We can find elements of a theory of social practices in the work of a multitude of social theorists in the last third of the twentieth century who are of diverse theoretical origin: Pierre Bourdieu has explicitly pursued the project of a 'praxeology' since Outline of a Theory of

Practice (1972) up to his latest Cartesian Meditations (1997). . . . Anthony Giddens (1979, 1984) develops his version of practice theory in the framework of a 'theory of structuration,' heavily influenced by late Wittgenstein. Michel Foucault . . . arrives in his late works on ancient ethics (1984a, b) at a framework of analysing the relations between bodies, agency, knowledge and understanding that can likewise be understood as 'praxeological.' In empirical sociology, cultural studies and anthropology it is above all works in the wake of Harold Garfinkel's ethnomethodology (1967), Judith Butler's 'performative' gender studies (1990) and Bruno Latour's science studies (1991) that can be understood as members of the praxeological family of theories" (Reckwitz 2002, 243).

8. Ray McDermott's 1996 work on the "acquisition of a child by a learning disability" in the edited volume *Understanding Practice* is a wonderful example of how labeled populations are made in practice, at the practical intersection of expert discourse, labels, institutions, children, and adults, all in relation to each other.

9. Notably absent from this list is religious tension. Sierra Leone's Christians and Muslims coexist quite peacefully.

10. Another popular way of dichotomizing analyses of the war is "greed or grievance," the idea that economic considerations are a better explanation for civil war than intergroup hatreds (Berdal and Keen 1997; Berdal and Malone 2000; Duffield 2000). The argument has been applied particularly often to "resource conflicts" and especially those fought around "conflict diamonds." Ian Smillie, Lansana Gberie, and Ralph Hazleton (2000), for example, claim that the war in Sierra Leone should be seen as primarily about clandestine diamond mining and the consequent ability of "warlords" to buy guns and other necessities of war.

11. This group of Sierra Leonean scholars published a collection of essays in *Africa Development* in 1997 and those essays, with a few additions, have been re-released as a book (Abdullah 2004a).

12. See also Zack-Williams (1999), Riley (1996 and 1997), and Riley and Sesay (1995). These scholars point to the corruption of the preceding APC regime, the impact of structural adjustment policies, and the side effects of the Liberian war.

13. See my work on youth style in Sierra Leone (Shepler 2004).

14. Among most ethnic groups in Sierra Leone, all men join the Poro society or a similar men's society. For example, the primary male society of the Limba and Loko in the north is Gbangbani. Women join the Sande or Bondo society. These societies, though secret, have been widely studied, especially their masked dances (Little 1967; MacCormack 1979; Nunley 1987; Murphy 1980; Bledsoe 1984). The societies operate as an important site of political power, making decisions about how laws are carried out, as well as the initiations. Because they are secret, nonmembers do not know a lot about many of their specific practices. One of their primary functions is to promote solidarity within gender groups and age groups. Secret societies are a political institution, and secrecy functions

as a form of political power (Ferme 1999). Rosalind Shaw has speculated that the specific nature of the institution, in particular the reliance on secrecy, may have grown out of centuries' old uncertainties of the slave trade (2002).

In addition to the two main societies that everyone is supposed to join, there are smaller societies organized around specific occupations and the control of the mystical forces needed to perform the occupation (see, for example McNaughton [1988], on the society of Mande blacksmiths.)

15. *Kamajoisia* in Mende. Sometimes also transcribed *Kamajors*. I will call them Kamajohs since that is how they are referred to in Krio and in Sierra Leonean national media.

16. Although the CDF is primarily associated with the ethnically Mende Kamajohs, the first use of traditional hunters to fight the rebels started in the north among the Kuranko ethnic group. Their fighters were known as Tamaboros. The idea spread to the south and the Kamajohs were formed. See Muana (1997) for an in-depth history of the formation of the Kamajohs and their place in the politics of the nation. See Ferme and Hoffman (2004) and Leach (2000) for more on the figure of "the hunter" in the organization of the force.

17. My ethnographic sensibilities align with several scholars of postconflict West Africa, most notably Coulter (2009), Utas (2003, 2004, 2005a, 2005b, 2005c), Vigh (2006), and Peters (2006).

18. To help select field sites, I consulted with various child protection NGOs about possible places to do the research I had planned. I was restricted by the security concerns of the UN peacekeepers, who did not want me to work in rebel-controlled territory.

19. One problem with my sample is that the most rural areas I studied were also Temne areas, and the Mende areas I worked in were more urban or unusual (such as a refugee camp). I could have also studied a small Mende village or a large Temne town, but time and safety did not permit it. Now that all areas of the country are accessible, it would be useful to conduct similar research in the east (especially in Kono areas) and in the far north among Limba and Kuranko.

NOTES TO CHAPTER 1

1. I certainly do not mean to imply that there is only one version of Western childhood. In the West, childhood is also a contingent category, struggled over, made and remade in multiple ways.

2. The very earliest work in the field was Cohn and Goodwin-Gill 1994; see also Legrand 1997, 1999; Bennett, Gamba, and van der Merwe 2000; de Berry 2001; Coalition to Stop the Use of Child Soldiers 2002, 2003. Then we see some refinement of focus with the work of Peter Singer (2005), whose focus was on how the U.S. military should respond to armies made up of child soldiers, journalistic accounts by Jimmie Briggs (2005), and work by David Rosen (2005) that uses a comparative study to point out the contradictions inherent in the modern focus on child soldiering.

A series of works by child protection practitioners (Honwana 2006; Wessells 2007; Boothby, Strang, and Wessells 2006) followed, building on successful programs around the world, and integrating some of the findings from early anthropological work. Myriam Denov's work (2010) is based on participatory research and focus groups, that of Theresa Betancourt and colleagues (Betancourt, Pochan, and de la Soudiere 2005; Betancourt et al. 2008; Betancourt, Agnew-Blais et al. 2010) is based on psychological indicators. Susan McKay and Dyan Mazurana's work, together and separately, is based on interviews (McKay 1998, 2000; Mazurana and McKay 2001; Mazurana et al. 2002; McKay 2004; McKay et al. 2004; McKay and Mazurana 2004; McKay 2006). McKay has recently embraced participatory action research as a method (see McKay et al. 2010).

In Africa there has been country-specific work on child soldiers in Mozambique (Gibbs 1994; Thompson 1999; Honwana 1999; West 2000; Schafer 2004; Charnley 2006); Angola (Wessells and Monteiro 2000; Honwana 2001); Uganda (Ehrenreich 1998; de Berry 2004; Corbin 2008; Annan, Brier, and Aryemo 2009); Liberia (David-Toweh 1998; Peters 2000; Utas 2003; Ellis 2003); and Sierra Leone (Zack-Williams 2001; de la Soudiere 2002; Shepler 2003; Krech 2003; Shepler 2004; Williamson 2005; Peters 2007; Denov 2010).

3. There is a whole strand of work on war-affected youth that emphasizes their agency, as a way of counteracting their presumed status as helpless victims. It is an important corrective to the human rights discourse, but in my view it does not open up productive avenues for further understanding, policy making, or program design.

4. See Shaw (2002) for more on the long-lasting cultural impacts of slavery on the culture of the region.

5. In Sierra Leone, infants or toddlers "*no get sense*" but, as Ferme (2001b, chapter 6) and Gottlieb (1998) argue, they nevertheless have a powerful social meaning because of the perceived relationship of newborns with the world of the spirits. A child, or *pikin*, "*get sense*," can understand, and talk.

6. The same woman told me she thought that the focus on child soldiers was a totally externally driven phenomenon. "Most Sierra Leoneans, if they were honest, would say to kill them all."

7. It is not surprising from a Marxist perspective that relations of youth to adult, as is the case with many other social relations, are related to the mode of production.

8. Urban definitions of childhood are also different from rural ones. One day in Freetown, I was walking with M. T., an old friend of mine from my days as a Peace Corps teacher. He and I would often discuss the differences between up-country life and Freetown life. On our way to his neighborhood bar, we saw two men on a motorcycle, the man on the back carrying a small bundle wrapped in white cloth. M. T. said to me, "You know what that is, right? It's a dead baby. They are taking it to the burial ground." Since we had been talking

about models of youth, I asked more about child burials. He told me that rural people found it odd how seriously Freetown people took the death of children, sometimes even placing death announcements in the newspaper. In his wonderful ethnography of the Kuranko of Sierra Leone, Michael Jackson states with respect to infant burial, "People are enjoined not to express grief or mourn at an infant's death since 'tears burn the child's skin and cause it pain.' The actual burial of a child is a perfunctory affair involving only the immediate family. Some consolation may be offered in the belief that the infant once dead may be reborn. . . . Infant dead are buried at the back of the house, in the domestic area . . . where women prepare food and cook, where domestic refuse is discarded" (1989, 75).

9. Chris Coulter (2009) and Myriam Denov (2010) make similar points about "structural continuity" before, during, and after the war.

10. Esther Goody notes that the placing of children outside the natal family of orientation is common in rural areas of Sierra Leone (1982, 210). It has been reported for the Mende, the Limba, the Temne, the Fula, and the Sherbro.

11. This type of fostering is known as *Kafala* in Islam, and is a requirement for Muslims. In *Kafala* a child may be placed under the guardianship of a family, but the child continues to retain his or her lineage. See also (Mattar 2003). *Kafala* is mentioned explicitly in the UN's Convention on the Rights of the Child.

12. Many seminars and many years of study with Professor Jean Lave have helped form my ideas on this topic.

13. Murphy finds the patron-client relationships between commanders and child soldiers a good explanation for youth participation in war, stating, "many youths became dependent on the patronage of military commanders as a way to transform their physical vulnerability and economic desperation. Patronage also provided them with a response to the political marginalization and economic destitution enforced by the corrupt regimes of the nation state. As a phenomenon, therefore, child soldiers illustrate a broader principle of youth clientelism in Africa (and elsewhere): the social production of dependency on patronage when local and national structures fail to provide for the social and economic needs of youth" (Murphy 2003, 62)

14. Because of the importance of these secret associations, especially the men's Poro, the tribes of the region are sometimes know as 'the Poro tribes' or 'Poro cluster', including "the Lokko, Temne, Kono, Mende, Bullom, Krim, and Sherbro of Sierra Leone and, in Liberia, the Gola, Vai, De, Kpelle, Kissi, Gbande, Belle, Loma, Mano and Gio" (d'Azevedo 1959, 68).

15. The situation is more complicated for the Krio. There are Krio-only societies that serve roles similar to those of indigenous secret societies. Also, some Krio have chosen to join secret societies for political or other reasons. In addition, the rise of evangelical Christianity has meant that some people refuse for their children to join secret societies for religious reasons.

16. Michael Jackson explains that for the Kuranko (the ethnic group in Sierra Leone he knows best), and for several other West African groups, "the contrast between bush and town signifies the extremes between exuberant disorder and social order, or between uncontrolled power and restraint. Because the bush is a source of vital and regenerative energy, the village must open itself up perennially to it. Farmers clear-cut the forest in order to grow rice that is the staff of life. Hunters venture into the bush at night, braving real and imagined dangers in their search for meat" (2004, 156).

17. Thanks to M. T. Bangura for help translating interviews that were taped in a mixture of Krio and Temne.

18. For boys, this means the removal of the foreskin. For girls, it means the removal of the clitoris and labia minora, with the whole area then sown up. This is also known as female genital mutilation in human rights discourse.

19. Stephen Ellis also sees the connection, noting, "Other militias, although not an emanation from historically existing secret societies, nonetheless function in a mode of initiation, such as the National Patriotic Front of Liberia (NPFL) and the Revolutionary United Front (RUF), both of which have initiated newcomers with rituals, tattoos and scarification that resemble traditional techniques to some extent" (2003, 4).

20. Immediately after the war, UNICEF oversaw a project to remove the scars from children to help ease their reintegration.

21. We can see both models operating in present day Africa. The "young lions" of the victorious anti-apartheid movement in South Africa are understood much differently than the defeated young fighters of Sierra Leone.

22. See Nathaniel King's 2012 doctoral dissertation on the same societies in the postwar period.

23. Stathis Kalyvas (2006) makes a related claim about the importance of local cleavages to action on the ground in civil wars.

24. Of course these macrolevel factors complement microlevel factors, such as the need of individual commanders for additional fighters.

NOTES TO CHAPTER 2

1. "Animator" was the international NGO's vocabulary for the ICC staff. When I said I had never heard it before, they explained to me, "it means something like teacher." Robert Krech (2003) believes the term comes from *Training for Transformation* by Anne Hope and Sally Timmell, a popular Freire-influenced guide to running community workshops.

2. Kenema is the capital of the Eastern Province.

3. Mine is not a description of the system as a practitioner or policy maker would give it, nor am I attempting a program evaluation. Fine technical descriptions of the whole DDR system for children, now part of the "lessons learned" documentation produced by and for child protection experts, are given by Andy Brooks (2005), UNICEF child protection officer in Sierra Leone and one of the

key architects of the system, and by John Williamson (2006), a social worker and USAID staffer.

4. On March 30, 2001, UNAMSIL was authorized to have a maximum strength of 17,500, making it the largest UN peacekeeping operation in the world up to that point (http://www.un.org/Depts/dpko/missions/unamsil/index.html).

5. There has been some excellent scholarly work on these transitional justice initiatives. See Shaw (2005), M. Kelsall and Stepakoff (2007), and T. Kelsall (2009).

6. In 2002 the Government of Sierra Leone, in an effort to guide and supervise all of this NGO work on child protection, established a National Commission for War Affected Children, but it has had minimal impact.

7. See Brooks (2005, 10–13) for detailed information about struggles over categorization of ex-combatants as children or adults at demobilization. I discuss demobilization in greater detail in chapter 4, which focuses on the differences between formal and "spontaneous" reintegrators.

8. In 2006, well after my fieldwork, the Inter-Agency Disarmament, Demobilization, and Reintegration Working Group developed a set of guiding principles on the subject of "Children and DDR." These principles outlined five programming-related areas for effective reintegration:

psychosocial support and care

community acceptance

education, training, and livelihoods

inclusive programming for all war-affected children

follow-up and monitoring (quoted in Stark, Boothby, and Ager 2009, 525).

9. The Makeni ICC was not always in operation, depending on the state of the conflict, and Daru was deemed too dangerous for me to visit.

10. About the field sites described in the introduction; Pujehun is in the Southern Province so was served by the Bo ICC, a day's journey to the north; Masakane is in the Northern Province so was served by one of Caritas Makeni's ICCs in either Port Loko, Lungi, or Makeni; Rogbom was close to Freetown, but still technically in the Northern Province, so it was served by Caritas Makeni as well as by the Lakka ICC outside Freetown; the Jerihun camp was mainly inhabited by people originally from the east, so they were served by the IRC ICC in Kenema, the capital of the Eastern Province.

11. *The Harmony of Illusions: Inventing Post-Traumatic Stress Disorder* by Allan Young (1995) is very revealing about the social forces involved in getting PTSD originally recognized as a legitimate disorder. Since its inclusion in the DSM (American Psychiatric Association's *Diagnostic and Statistical Manual of Mental Disorders*), its use as a diagnosis has spread far beyond the Vietnam War veterans who were its first acknowledged victims.

Patrick Bracken, in "Deconstructing Post-Traumatic Stress Disorder" (1998), situates PTSD in the history of its creation in Western psychology and questions the relevance of Western forms of therapy to non-Western societies. He claims the discourse on trauma has been largely shaped by cognitivism. A central tenet of this approach, the need for successful "processing" of a traumatic experience, is now widely accepted. Cognitivism involves a strongly individualistic approach, universality of the forms of mental disorder, and the relevance of Western therapy in non-Western settings.

12. Although, interestingly, in the United States we are getting tougher and tougher on child criminals. For some reason though, in the West, Sierra Leonean child soldiers are still mainly understood as victims. See Fofanah (2004) for work on shifting notions of juvenile justice in Sierra Leone.

13. The same lament occurs in many media reports on the problems of child soldiers. See, in particular, Sorius Samura's *Cry Freetown*, which aired on CNN in February 2000.

14. Melissa Leach agrees, saying "Foreign-development agencies and donors have seized on and are investing heavily in hunters (Kamajohs in Sierra Leone) as the embodiment of the 'authentic African culture' so valued in contemporary development discourses" (2004, x).

15. Mike Wessells agrees, noting that "when community-based practitioners such as those of IRC or CCF (now ChildFund) supported traditional ceremonies, it was not for the purposes of detraumatizing or de-initiation but for making harmony with the spirits" (personal communication, January 2012, Wessells to Shepler). He says he met numerous healers in the Northern Province of Sierra Leone who spoke of the importance of traditional ceremonies as a means of making harmony with the spirits. See also Stark, Ager, Wessells, and Boothby (2009) for examples of the usefulness of ceremonies to girls in Sierra Leone.

16. Tim Allen writes perceptively about the use and abuse of the "traditional" *mato oput* ritual in northern Uganda, arguing that "the merits of reifying local rituals in a form of semi-official 'traditional justice' have been oversold and the dangers underappreciated" (2008, 47).

17. UNICEF reported in May 2002 that more than 6,760 children were benefiting from the CREPS program. In 2003, the program's most successful year, CREPS programs enrolled more than 11,209 children (Wang 2007, 39).

18. According to the Women's Commission for Refugee Women and Children (2004), organizations include ActionAid, the American Refugee Committee (ARC), CARE, Catholic Relief Services (CRS), Children in Crisis, Christian Children's Fund (CCF), Concern, DFID, The European Union, the Forum for African Women Educationalists (FAWE), German Agency for Technical Cooperation (GTZ), International Rescue Committee (IRC), Islamic Development Bank, Management Systems International, the Norwegian Refugee Council (NRC), Plan International, USAID, World Relief, and World Vision.

19. Krech also notes that at demobilization children were given only two options: either education or skills training (2003, 140).

20. Gara tie dyeing is a West African method of dyeing fabric. It originated north of Sierra Leone, and was originally done with indigo dyes. Elaborate patterns are tied into the fabric before it is dyed. Other techniques include wax block, flour

paste, and batik. Before the war, it was a skill mostly practiced by women in the north. The northern capital, Makeni, was particularly well known for its fine gara dyers.

21. Krech reports that some NGO staff found that when they participated in interviewing demobilizing child combatants, the format was consistently the same. Children were asked to form a line and in turn pick a skill from the list. The problem was that if the first child said "carpentry," the second child in line behind also said "carpentry" because he wanted to be with his friend. As a result NGOs sometimes revisited villages and did their own needs assessments based on what the community said were in demand skills.

22. Coulter draws many of the same conclusions about skills training for girls and young women. She notes that although an NGO director she spoke with told her that the programs were designed by "ex-pats in Freetown," asking around for a few days in the wider community yielded quite a number of alternative skills training activities (Coulter 2009, 189).

Since the period of my fieldwork, and in part in response to critics of the skills-training programs, new approaches are in place in many other post-conflict contexts. Drawing on lessons learned from the microcredit world, practitioners are focusing more on market assessments to drive the choice of skills training on offer, and even including youth in the market assessments for stronger workplace development programs (Beauvy-Sany et al. 2009).

23. Krio for "there is no place to throw away a bad child." This is a well-known saying in Krio about the necessity of accepting even bad children.

NOTES TO CHAPTER 3

1. Another example comes from a visit I made to the Port Loko ICC with my friend Wusu. He was curious to see the ICCs I had been talking about, and an old friend of his was on the staff, so he came with me to Port Loko for a week. As I have said, it was surprisingly easy for people to move in and out of the ICCs, and Wusu spent the first day of my visit in the ICC talking to the children and to the staff. One day was enough for him, so when I went back the following day, he planned to drop me off and set off on his own. But first, he took aside one smallish boy we had met the day before. "Do you know me?" he asked. The boy sheepishly responded, "Yes sir." Wusu replied, "You know, the family is all very worried about you. Of course I will have to tell them where you are." "Yes sir" the boy mumbled. It turned out the boy was not a child soldier at all, and in fact knew very well where his family was located. He was a boy attached to the household of Wusu's uncle in Makeni. The boy, according to Wusu, had decided he did not want to work and so had gone off on an adventure. Wusu later reported the boy's whereabouts to his guardians and they went to pick him up.

2. It is interesting that although UNICEF and others would talk about drama activities as giving children a way to reconnect to "traditional" forms of

expression, the drama teacher was an educated Krio from Freetown whose class background was much different from the children at the ICCs. The language of the skit was not the children's everyday language.

3. Another more recent psychosocial, dance-based activity for former child soldiers is detailed in Harris (2010).

4. Mats Utas in his dissertation work in Liberia and Sierra Leone, and in further work since, uses the term "victimcy" to describe this phenomenon. See (Utas 2003, 2005a, 2005b).

5. That is, clothes in African styles made from locally dyed fabrics or cotton imported from other nations in Africa. "*Lehk Yu Culture*" ("Like Your Culture") was a popular song during my fieldwork, and the wearing of "Africana" clothes was seen as patriotic.

6. Sengbe Pieh is the local name of the Sierra Leonean man at the center of the Amistad story. Even before the war, the school had benefited from some aid from the United States in recognition of this historical tie.

7. On one occasion I observed her home economics class. She dictated notes about how to select the right furniture for your home. She said to polish the furniture with "good quality furniture cream," further commenting, "You are very unfortunate that you will never see such a thing. You will only read about it on paper."

8. See my early work on education as a site of political struggle over futures in Sierra Leone (Shepler 1998).

9. See Brooks (2005) for details of this event from UNICEF Sierra Leone's perspective.

NOTES TO CHAPTER 4

1. While Pa Kamara was telling me this story, he pointed out a young man passing by and told me that he had joined the rebels while they were occupying the area. When the young man thought Pa Kamara was dead, he went to his house and took all his belongings. Later, Pa Kamara saw all his things for sale in the boy's hand. When the facts were revealed, the young man's family could only beg for forgiveness.

2. Following UNICEF's "Paris Principles" terminology, "child soldier" means any person under 18 years of age who is part of any kind of regular or irregular armed force or armed group in any capacity, including but not limited to cooks, porters, messengers, and those accompanying such groups, other than purely as family members. It includes girls recruited for sexual purposes and forced marriage. It does not, therefore, only refer to a child who is carrying or has carried arms (UNICEF 2007).

3. David-Toweh (1998) reports that in neighboring Liberia, only about one-third of child combatants went through the formal demobilization process.

4. As reported by Jon Pederson, "in general, what most studies supply is analysis of the children enrolled in the programs. We seldom learn about those that did

not get enrolled, or those who dropped out. This selection effect problematic is similar to that which plagues any study of rehabilitation programs, vocational training programs, affirmative action programs or similar. It is well known that disregarding selection effects may lead to biased and misleading results. Therefore, more formal design of such studies would be a useful contribution. Similarly, studies of recruitment into soldiering that take account of selection effects would be useful in order to identify factors that may be changed in order to reduce recruitment" (2001, 13).

5. These figures come from NGOs such as Radda Barnen and the Coalition to Stop the Use of Child Soldiers. McKay and Mazurana repeat those figures and report that their field data confirm a high percentage of girls within the RUF and AFRC. "The majority of the females and males in the respective forces interviewed responded that children comprised approximately half of the RUF and AFRC forces in camps or compounds where they were held, and all reported the presence of girls in numbers equal to or slightly less than boys" (2004, 91). They go on to estimate that around 33 percent of the children in the SLA were girls and 10 percent of the children in the CDFs were girls. Based on my field data, these last numbers seem high. In particular, I never heard of any girls in any CDFs. I will have more to say about this issue in chapter 5.

6. The Survey of War Affected Youth in Uganda (SWAY) is an example of a way of addressing the gap between large-scale but superficial and small-scale but deep approaches. The researchers carried out a random survey of a large population of youth in northern Uganda and from there determined what portion of the population was war-affected, and indeed what portion were so-called spontaneous reintegrators. A large survey method like theirs can be a useful complement to in-depth ethnographic work. For much more on this project see the website: www.SWAY-Uganda.org.

7. Partly because I was often mistaken for an NGO worker, and people thought I might be able to help them get registered.

8. Theresa Betancourt's multiple publications on a series of follow-up psychological screenings with a set of former child soldiers yield interesting, and somewhat contradictory data on this issue as well (Betancourt, Pochan, and de la Soudiere 2005; Betancourt, Agnew-Blais et al. 2010; Betancourt, Borisova et al. 2010). In the earlier study for the IRC, there was a relatively small difference in psychosocial measures between ex-RUF by NGO presence, and that small difference might be explained by the fact that the non-NGO sample were all in one location different from those in the NGO sample. In her later work, she is looking for risk factors and resilience factors and somehow never reports whether there are differential outcomes for formal and informal reintegrators.

9. *San san* means "sand" in Krio, and *san san* boys are youth who work in the diamond-mining areas sifting sand to hunt for diamonds. "*San san* boys" is

one of the terms for a specific youth culture (along with *savisman, dregman, ngiahungbia ngorngesia*) suggested for analysis by the Sierra Leonean scholars of Africa Development (Abdullah et al. 1997).

10. Throughout my fieldwork, I tried to stay as far away from men with guns as possible, but once I stumbled right into a demobilization exercise. I was on the way to the ICC at Port Loko in an NGO vehicle when we saw a commotion on the road ahead. The driver put on the brakes as soon as he saw the soldiers, but they beckoned us forward. There were probably two hundred to three hundred fighters of various stripes hanging around a small village along the road. There were a few Indian UNAMSIL officers, some other soldiers, and some who were obviously rebels. The other people riding in the vehicle with me were frightened, and their fear affected me. Almost every man there had a gun. I saw some rocket-propelled grenade launchers (RPGs), but mostly AK-47s. Some were carrying bags of ammunition. We slowly drove through the village and continued on our journey.

11. See Hoffman (2003) for a description of a CDF demobilization exercise in Bo.

12. See UNICEF/NCDDR Guidelines for Assisting Children from the Fighting Forces in the DDR, October 2000; UNAMSIL Procedures for processing Children Through the DDR Programme, April 2000; NCDDR Phase III Joint Operational Plan (JOP), December 2000.

13. It is no coincidence that the flow of children into the program peaked in October and November 1999 and trailed off up until the outbreak of hostilities in May 2000. In a sample study done by NCDDR, the entry of child soldiers dipped from 31 percent of the overall population of demobilizing combatants in October to less than 7 percent in April (Brooks 2005).

14. Commonly known as "CP Com," these meetings were held monthly at the Ministry of Social Welfare, Gender, and Children's Affairs. UNICEF and NGO representatives attended. These meetings were a primary site for reporting statistics and national policy discussions.

15. This is hardly a new idea. Akhil Gupta (2001), for example, points to enumeration as key to the functioning of governmentality.

16. "Informants claimed that both RUF and CDF commanders collected weapons from their fighters and redistributed them to kin or friends, or traded weapons on the black market to allow purchasers to register dependents as 'child combatants' (to access free basic education offered on demobilization.) For every false ex-combatant there must be a real ex-combatant without benefits. The numbers excluded in this way are in dispute, . . . but various ex-combatant sources put the number (excluded in this way) as high as 50-60 percent of all gun-carrying ex-combatants" (Richards et al. 2003, 4).

17. In Mazurana and Carlson's sample of girls who did not go through DDR (N=25) 46 percent cited not having a weapon that was required for entry as their reason for not formally demobilizing. They also found that 100 percent of their study

population who entered DDR (N=25) were asked to turn in a weapon and perform a weapons test (McKay and Mazurana 2004, 100).

18. This is especially true for girls; see chapter 5.

19. In Sierra Leone, to have a cool heart is to be calm and reasonable. To have a
warm heart is to be quick to anger, or out of control.

20. When I was looking for a field site to study children who had been reunified
for a long time, I was a little disappointed at their dearth. I looked through
the records at Lakka for cases of children who had been reunified before 1997.
The record keeping was awful. When I asked the priest in charge for help, he
explained that sometimes early on they did not keep records for political or
safety reasons. Sometimes an RUF commander would free a boy against the
wishes of his higher-ups and report that the boy had escaped, so they had to
be careful about admitting any knowledge. One point of this story is that early
on, programs were more sensitive to the need for secrecy or at least secrecy as a
reintegration strategy. That has become less and less the case with the increasing importance of the formal reintegration system. In other words, the form of
the child protection system shifted over time as the nature of the war shifted,
and children who were involved in the system at an earlier point therefore have
different meanings of "child soldier" to contend with.

21. "RUF groups received some local support from civilians. Losers in a local land
or chieftaincy dispute might sometimes side with the insurgents to secure
revenge. . . . Local support for the RUF may have been strongest in Pujehun
District because of the mid-1980s Ndogboyosoi movement" (Richards 1996, 8).

22. I want to be careful not to put too much of a tribal spin on things, as that over-
simplifies the issues, but it is certainly the case that many in Sierra Leone see the
conflict in precisely these terms.

23. And I would argue that even in formal programs, the more they relied on the
mechanisms of informal reintegration, for example, fostering children out of
ICCs, the more successful they were at reintegration. See chapter 4.

24. I met a man there working to renovate the chemistry lab. He said that UNESCO
was providing science lab equipment to schools throughout the country, but
since the schools lacked even the most basic provisions, it was like dumping the
science equipment directly into the marketplace.

25. I had also asked Father Momoh, the head of CAW, about it earlier. He told
me CEIP was administered by IRC in the south and that they had submitted a
list of their beneficiaries with DDR numbers to IRC so they had done all they
could do.

26. Richards and colleagues discuss the nature of "the community" in rural Sierra
Leone, and analyze the main sources of poverty and vulnerability. They argue
that women, youth, and "strangers" have been politically marginalized, and that
the rural community is typically divided between the leading lineages and the
rest (Richards, Bah, and Vincent 2004, 40).

27. There were similar lists of adult ex-combatants organized to get the most out of adult demobilization, disarmament, and reintegration programming. Ferme and Hoffman report that "the determination of who qualified as a combatant became the prerogative of a few highly influential people in the (CDF). In many cases, these determinations were based not on actual field experience, but on the willingness to pass on to a CDF patron a percentage of the commodities and financial inducements given to ex-combatants" (2004, 87). Fanthorpe and Maconachie (2010) discuss a more recent example of a former combatant hoping to attract donor funding to get his NGO off the ground, keeping up a local network of ex-combatants and facing competition from other NGOs with stronger community links.

28. Richards, Bah, and Vincent include a lengthy discussion of the rise of the "Village Development Committee" in rural areas (2004, 24–25).

29. Pa Kamara, the headman of Rogbom, told me a story of a local man named Anthony. He came around and everyone contributed money for him to go to NGOs and try to get help. He disappeared, and the community has seen nothing from him.

30. Robert Krech tells a related story about communities in Sierra Leone leveraging NGOs against each other for a school reconstruction project (2003, 151).

31. There was anger about the government's NCRRR (National Commission for Rehabilitation, Resettlement, and Reconstruction) signboard on the main road near the village saying that NCRRR is helping to rebuild the school. "They've done nothing, but they get to post their signboard right on the main road." There was concern that other NGOs would not help if they saw that this area was taken.

32. See *Participation: The New Tyranny?* (Cooke and Kothari 2001) on the development industry's recent reliance on the forms of community participation as a panacea. See also "The Paradoxes of Community-Based Participation in Dar es Salaam" for a skillful demonstration of how "the channeling of participation into CBOs has, overall, narrowed the participatory opportunities available to individuals, the agendas they pursue and the preferences they communicate to government officials and other development actors" (Dill 2009, 739).

33. However, one woman came around and complained that she did not think she was included among the beneficiaries since she had heard the program was only for the parents of children who had been taken by the rebels. She begged the headmaster to include her name on the list anyway.

34. PLAN is an international child protection NGO that operated in Sierra Leone even before the war. Sadly for PLAN, *"plan"* has only bad connotations in Krio.

NOTES TO CHAPTER 5

1. I see this intellectual project as growing out of the work of historical anthropologists of colonialism such as Jean Comaroff and John Comaroff and Ann Stoler (Comaroff and Comaroff 1991; Comaroff and Comaroff 1997; Stoler 1995).

2. "Techniques of mystical warfare . . . became incorporated into the fighting forces of the state itself. In the early 1990s, a reason commonly given by both soldiers and civilians for the army's failure to defeat the RUF rebels was that the rebels' expertise in medicines and rituals of defense was superior to that of the Sierra Leone Army. The RUF used (and sometimes abducted) Muslim ritual specialists—*mori men* in Krio—to make amulets and other ritual materials" (Shaw 2003, 90).

3. See Besteman (1996) for a related exploration of the "othering" of Somalia and the Somalian conflict in the Western media.

4. The initiation into the Kamajohs, and other hunting societies, involves being washed with water in which a specific leaf has been soaked. Different leaves are understood to have different properties.

5. I also found out about a peculiar kind of child combatant, the *bao tchie*, a young boy around three years of age, who was supposed to have special powers. The adults would bring him to the front lines for protection.

6. Notice that the supposedly modern rebels were assumed to be using traditional magical methods to hide themselves.

7. Recall that the Kamajohs are the ethnically Mende branch of the CDF. Ferme and Hoffman (2004) are only discussing Kamajohs, whereas I am discussing the CDF of different ethnicities.

8. This is one possible reason why Kamajohs eventually started using some of the same tactics as the RUF at checkpoints—shaking down drivers for money and goods.

9. See Shepler (2010c) on oppositional youth culture expressed through music around the 2007 elections.

10. See, for example, the work of Christine Liddell, Jennifer Kemp, and Molly Moema. (1993) on the "young lions" of South Africa.

11. Macartan Humphreys and Jeremy Weinstein (2004b) report the same kind of dissatisfaction among adult CDF members and note that in the postwar period ex-combatants of all factions are realizing that they have a great deal in common and are uniting against the government agency for ex-combatants (NCDDR).

12. McKay and Mazurana (2004) report large numbers of girls and women in the CDF. This contradicts what all my informants told me. I think they must be talking about women and girls who served very important support roles without actually being initiated into the society. According to Hoffman (personal communication) there was a woman in Bo who carried out the initiations, but she was not a combatant.

13. Rogbom was held by rebels (RUF, AFRC, and West Side Boys) after they were driven from Freetown after the January 6th attack.

14. Brooks (2005), assessing the formal programs for child ex-combatants for UNICEF, reports that as the girl combatants moved into the interim care centers and reintegration phase, there was a lack of attention to the need

for separate and gender-specific services for girls. He concludes that pro-gramming was shaped by the profile of beneficiaries, who tended to be male adolescents.

15. Exceptions are Coulter's ethnographic work (2009), already mentioned; MacK-enzie's interviews with women ex-combatants in Makeni (MacKenzie 2009a, 2009b); Stark and colleagues' work on developing culturally appropriate indica-tors for measuring girls' reintegration (Stark et al. 2009); and McKay and col-leagues' participatory action research with war-affected young mothers (McKay et al. 2010). See also an excellent piece of research on girls abducted by fighting forces in Angola by Vivi Stavrou (2004).

16. See Baldi and MacKenzie (2007) for more on the issues facing the children of these young mothers.

17. This strategy is not only suggested for ex-combatants, the National Forum for Human Rights reports that "in some parts of the country, perpetrators of rape are encouraged to marry the victim" (2001, 12).

18. Caroline Bledsoe cites a Sierra Leonean secondary school teacher who says, "The schools in Sierra Leone do not generally admit girls who have given birth: mothers. She is not considered a school girl again" (1990c, 293).

NOTES TO THE CONCLUSION

1. William Murphy links this discussion to "non-refoulement" for refugees (Mur-phy 2010).

2. After some early work by Peter Singer (2005), whose focus was on how the U.S. military should respond to armies made up of child soldiers, and journalistic accounts by Jimmie Briggs (2005), David Rosen (2005) made a comparative study to point out the contradictions inherent in the modern focus on child soldiering. A series of works by practitioners (Honwana 2006; Wessells 2007; Boothby, Strang, and Wessells 2006) followed, building on successful programs around the world and integrating some of the findings from early anthropologi-cal work. Myriam Denov (2010) is based on participatory research and focus groups; the work of Betancourt and colleagues (Betancourt, Pochan, and de la Soudiere 2005; Betancourt et al. 2008; Betancourt, Agnew-Blais et al. 2010) is based on psychological indicators; McKay and Mazurana's work, together and separately, is based on interviews (McKay 1998, 2000; Mazurana and McKay 2001; Mazurana et al. 2002; McKay 2004; McKay et al. 2004; McKay and Mazurana 2004; McKay 2006). McKay has recently embraced participatory action research as a method. See (McKay et al. 2010).

3. As Norman Denzin explains, "The politics and political economy of evidence is not a question of evidence or no evidence. It is rather a question of who has the power to control the definition of evidence, who defines the kinds of materials that count as evidence, who determines what methods best produce the best forms of evidence, whose criteria and standards are used to evaluate quality evidence?" (2009, 142).

4. Scholars often say they are doing multi-method research because they do interviews in addition to surveys, but when they include qualitative work it tends
to be in the form of illustrative vignettes to help humanize the (presumably
more reliable) quantitative results. The vignettes rarely threaten the underlying
frameworks.

5. As Vanessa Pupavac so cogently points out, in her article "Misanthropy without
Borders: The International Children's Rights Regime," "Underlying the imperative . . . to institutionalise children's rights is an implicit mistrust of their carers"
(2001, 100).

Abbink, Jon, and Ineke van Kessel, eds. 2005. *Vanguards or Vandals: Youth, Politics and Conflict in Africa.* Leiden and Boston: Brill.

Abdalla, Amr, Suleiman Hussein, and Susan Shepler. 2002. "Human Rights in Sierra Leone, a Research Report to Search for Common Ground." Washington, DC: Search for Common Ground.

Abdullah, I., Y. Bangura, C. Blake, L. Gberie, L. Johnson, K. Kallon, S. Kemokai, P. K. Muana, I. Rashid, and A. Zack-Williams. 1997. "Lumpen Youth Culture and Political Violence: Sierra Leoneans Debate the RUF and the Civil War." *Africa Development* 22 (3/4): 171–215.

Abdullah, Ibrahim. 1997. "Bush Path to Destruction: The Origin and Character of the Revolutionary United Front (RUF/SL)." *Africa Development* 22 (3/4): 45–76.

———. 2002. "Youth Culture and Rebellion: Understanding Sierra Leone's Wasted Decade." *Critical Arts* 16 (2): 19–37.

———. 2004a. *Between Democracy and Terror: The Sierra Leone Civil War.* Dakar: CODESRIA.

———. 2004b. "Introduction." In *Between Democracy and Terror: The Sierra Leone Civil War,* edited by Ibrahim Abdullah. Dakar: CODESRIA.

Alber, Erdmute. 2003. "Denying Biological Parenthood: Fosterage in Nothern Benin." *Ethnos* 68 (4): 487–506.

———. 2004. "'The Real Parents Are the Foster Parents': Social Parenthood among the Baatombu in Northern Benin." In *Cross Cultural Approaches to Adoption,* edited by Fiona Bowie, 33–47. London and New York: Routledge.

Allen, Tim. 2008. "Ritual (Ab)use?: Problems with Traditional Justice in Northern Uganda." In *Courting Conflict? Justice, Peace and the ICC in Africa,* edited by Nicholas Waddell and Phil Clark, 47–54. London: Royal African Institute.

Andreas, Peter, and Kelly M. Greenhill. 2010. "Introduction: The Politics of Numbers." In *Sex, Drugs, and Body Counts: The Politics of Numbers in Global Crime and Conflict,* edited by Peter Andreas and Kelly M. Greenhill, 1–22. Ithaca: Cornell University Press.

Annan, Jeannie, Christopher Blattman, and Roger Horton. 2006. "The State of Youth and Youth Protection in Northern Uganda: Findings from the Survey for War Affected Youth." UNICEF. Available online at http://www.SWAY-Uganda.org

Annan, Jeannie, Moriah Brier, and Filder Aryemo. 2009. "From 'Rebel' to 'Returnee': Daily Life and Reintegration for Young Soldiers in Northern Uganda." *Journal of Adolescent Research* 24 (6): 639–667.

Archibald, Steven, and Paul Richards. 2002. "Converts to Human Rights? Popular Debate About War and Justice in Rural Central Sierra Leone." *Africa* 72 (3): 339–367.

Ariès, Phillipe. 1962. *Centuries of Childhood*. New York: Alfred A. Knopf.

Bah, Khadija Alia. 1997. *Rural Women and Girls in the War in Sierra Leone*. London: Conciliation Resources.

Baldi, Giulia, and Megan MacKenzie. 2007. "Silent Identities: Children Born of War in Sierra Leone." In *Born of War: Protecting Children of Sexual Violence Survivors in Conflict Zones*, edited by R. Charli Carpenter, 78–93. Bloomfield, CT: Kumarian Press.

Bangura, Yusuf. 1997. "Understanding the Political and Cultural Dynamics of the Sierra Leone War: A Critique of Paul Richards' *Fighting for the Rain Forest*." *Africa Development* 22 (3/4): 117–148.

Beauvy-Sany, Melanie, Sita Conklin, Ann Hershkowitz, Radha Rajkotia, and Carrie Berg. 2009. "Technical Note: Guidelines and Experiences for Including Youth in Market Assessments for Stronger Youth Workforce Development Programs." In *Youth and Workforce Development PLP Technical Note*, edited by Stephanie Chen, Fiona Macaulay and Laura Meissner. Washington, DC: SEEP Network.

Behnam, Nicole. 2011. "Awkward Engagement: Friction, Translation, and Human Rights Education in Post-Conflict." Ph.D. diss., University of Pennsylvania.

Bennett, Elizabeth, Virginia Gamba, and Deirdre van der Merwe. 2000. *ACT Against Child Soldiers in Africa: A Reader*. Pretoria: Institute for Security Studies.

Berdal, Mats, and David Keen. 1997. "Violence and Economic Agendas in Civil Wars: Some Policy Implications." *Millennium: Journal of International Studies* 26 (3):795–818.

Berdal, Mats, and David Malone. 2000. *Greed and Grievance: Economic Agendas in Civil Wars*. Boulder, CO: Lynne Rienner.

Besteman, Catherine. 1996. "Representing Violence and "Othering" Somalia." *Cultural Anthropology* 11 (1): 120–133.

Betancourt, Theresa, Shawna Pochan, and Marie de le Soudiere. 2005. Psychosocial Adjustment and Social Reintegration of Child Ex-Soldiers in Sierra Leone. Unpublished report, IRC.

Betancourt, Theresa S., Jessica Agnew-Blais, Stephen E. Gilman, David R. Williams, and B. Heidi Ellis. 2010. "Past Horrors, Present Struggles: The Role of Stigma in the Association between War Experiences and Psychosocial Adjustment among Former Child Soldiers in Sierra Leone." *Social Science & Medicine* 70:17–26.

Betancourt, Theresa S., Stephanie Simmons, Ivelina Borisova, Stephanie E. Brewer, Uzo Iweala, and Marie de la Soudiere. 2008. "High Hopes, Grim Reality: Reintegration and the Education of Former Child Soldiers in Sierra Leone." *Comparative Education Review* 52 (4): 565–587.

Betancourt, Theresa Stichick, Ivelina Ivanova Borisova, Timothy Philip Williams, Robert T. Brennan, Theodore H. Whitfield, Marie De La Soudiere, John Williamson, and Stephen E. Gilman. 2010. "Sierra Leone's Former Child Soldiers: A Follow-Up Study of Psychosocial Adjustment and Community Reintegration." *Child Development* 81 (4): 1077–1095.

Bledsoe, Caroline. 1984. "The Political use of Sande ideology and symbolism." *American Ethnologist* 11 (3): 455–472.

———. 1990a. "'No Success without Struggle': Social Mobility and Hardship for Foster Children in Sierra Leone." *Man* 25: 70–88.

———. 1990b. "The Politics of Children: Fosterage and the Social Management of Fertility Among the Mende of Sierra Leone." In *Births and Power: Social Change and the Politics of Reproduction*, edited by W. Penn Handwerker, 81–100. Boulder: Westview Press.

———. 1990c. "School Fees and the Marriage Process for Mende girls in Sierra Leone." In *Beyond the Second Sex: New Directions in the Anthropology of Gender*, edited by Peggy Reeves Sanday and Ruth Gallagher Goodenough, 281–310. Philadelphia: University of Pennsylvania Press.

———. 1992. "The Cultural Transformation of Western Education in Sierra Leone." *Africa* 62, no. 2.: 182–201.

———. 1993. "Politics of Polygyny in Mende Education and Child Fosterage Transactions." In *Sex and gender hierarchies*, edited by Barbara Diane Miller, 170–192. Cambridge: Cambridge University Press.

Bledsoe, Caroline, and Uche Isiugo-Abanihe. 1989. "Strategies of Child Fosterage among Mende Grannies in Sierra Leone." In *Reproduction and Social Organization in Sub-Saharan Africa*, edited by Ron Lesthaeghe, 442–474. Berkeley: University of California Press.

Bøås, Morten, and Anne Hatløy. 2008. "Child Labour in West Africa: Different Work—Different Vulnerabilities." *International Migration* 46 (3): 3–25.

Boothby, Neil, J. Crawford, and J. Halperin. 2006. "Mozambique Child Soldier Life Outcome Study: Lessons Learned in Rehabilitation and Reintegration Efforts." *Global Public Health* 1 (1): 87–107.

Boothby, Neil, Alison Strang, and Michael Wessells. 2006. *A World Turned Upside Down: Social Ecological Approaches to Children in War Zones*. Bloomfield, CT: Kumarian Press.

Bourdieu, Pierre. 1977. *Outline of a Theory of Practice*. Cambridge and New York: Cambridge University Press.

———. 1997. *Méditations pascaliennes*. Paris: Seuil.

Boyden, Jo. 1997. "Childhood and the Policy Makers: A Comparative Perspective on the Globalization of Childhood." In *Constructing and Reconstructing Childhood: Contemporary Issues in the Sociological Study of Childhood*, edited by Allison James and Alan Prout, 190–229. London: Falmer Press.

———. 2000. "Social Healing in War-Affected and Displaced Children." Oxford: Refugee Studies Centre, University of Oxford.

———. 2001. "Conducting Research with War-Affected and Displaced Children: Ethics and Methods." Paper presented at Filling Knowledge Gaps: A Research Agenda on the Impact of Armed Conflict on Children, Florence, July 2–4.

Boyle, Elizabeth Heger. 2010. New child rights data: Implications for theory and research.

Bracken, Patrick. 1998. "Hidden Agendas: Deconstructing Post Traumatic Stress Disorder." In *Rethinking the Trauma of War*, edited by Patrick Bracken and Celia Petty, 38–59. London and New York: Free Association Books.

Briggs, Jimmie. 2005. *Innocents Lost: When Child Soldiers Go to War*. New York: Basic Books.

Brooks, Andy. 2005. *The Disarmament, Demobilisation, and Reintegration of Children Associated with the Fighting Forces: Lessons Learned in Sierra Leone, 1998–2002*. Dakar, Senegal: Imprimerie Graphi Plus.

Bucholtz, Mary. 2002. "Youth and Cultural Practice." *Annual Review of Anthropology* 31:525–552.

Burman, Erica. 1994. "Innocents Abroad: Western Fantasies of Childhood and the Iconography of Emergencies." *Disasters* 18 (3): 238–253.

Butler, Judith. 1990. *Gender Trouble*. London: Routledge.

Charnley, Helen. 2006. "The Sustainability of Substitute Family Care for Children Separated from Their Families by War: Evidence from Mozambique." *Children & Society* 20 (3): 223–234.

Christiansen, Catrine, Mats Utas, and Henrik E. Vigh. 2006. "Introduction." In *Navigating Youth, Generating Adulthood: Social Becoming in an African Context*, edited by Catrine Christiansen, Mats Utas and Henrik E. Vigh, 9–28. Uppsala: Nordic Africa Institute.

Clifton-Everest, Ian. 2005. "Meeting the Mental Health Needs of Children Who Have Been Associated with Fighting Forces. Some Lessons from Sierra Leone." In *Forced Migration And Mental Health: Rethinking the Care of Refugees and Displaced Persons*, edited by David Ingleby, 81–96. New York: Springer.

Coalition to Stop the Use of Child Soldiers. 2002. "Child Soldiers 1379 Report." London: Coalition to Stop the Use of Child Soldiers.

———. 2003. "Child Soldier Use 2003: A Briefing for the 4th UN Security Council Open Debate on Children and Armed Conflict." London: Coalition to Stop the Use of Child Soldiers.

Coalition to Stop the Use of Child Soldiers website, http://www.child-soldiers.org, accessed August 1, 2004

Cockburn, Cynthia. 2004. "The Continuum of Violence: A Gender Perspective on War and Peace." In *Sites of Violence: Gender and Conflict Zones*, edited by Wenona Giles and Jennier Hyndman, 24–44. Berkeley: University of California Press.

Cohn, Ilene, and Guy Goodwin-Gill. 1994. *Child Soldiers: The Role of Children in Armed Conflict*. Oxford: Clarendon Press.

Comaroff, Jean, and John Comaroff. 1991. *Of Revelation and Revolution: Christianity, Colonialism, and Consciousness in South Africa. Volume One*. Chicago: University of Chicago Press.

———. 2000. "Millennial Capitalism: First Thoughts on a Second Coming." *Public Culture* 12 (2): 291–343.

Comaroff, John, and Jean Comaroff. 1997. *Of Revelation and Revolution, Volume Two: The Dialectics of Modernity on a South African Frontier*. Chicago: University of Chicago Press.

Cooke, Bill, and Uma Kothari. 2001. *Participation: The New Tyranny?* London: Zed Books.

Corbin, Joanne N. 2008. "Returning Home: Resettlement of Formerly Abducted Children in Northern Uganda." *Disasters* 32 (2): 316–335.

Coulter, Chris. 2009. *Bush Wives and Girl Soldiers: Women's Lives Through Peace and War in Sierra Leone*. Ithaca, NY: Cornell University Press.

Cox, Herbert. 1956. Report of Commission of Inquiry into the Disturbances in the Provinces, November 1955–March 1956. Freetown, Sierra Leone.

d'Azevedo, W. L. 1959. "The Setting of Gola Society and Culture: Some Theoretical Implications of Variations in Time and Space." *Kroeber Anthropological Society Papers* 21:43–125.

Daniel, E. Valentine. 1996. *Charred Lullabies: Chapters in an Anthropology of Violence*. Princeton, NJ: Princeton University Press.

David-Toweh, Kelly. 1998. *The Disarmament, Demobilization, and Reintegration of Child Soldiers in Liberia: 1994–1997: The Process and Lessons Learned*. New York: UNICEF Liberia and the U.S. National Committee for UNICEF.

de Berry, Joanna. 2001. "Child Soldiers and the Convention on the Rights of the Child." *Annals of the American Academy of Political and Social Science* 575: 92–105.

———. 2004. "The Sexual Vulnerability of Adolescent Girls during Civil War in Teso, Uganda." In *Children and Youth on the Front Line: Ethnography, Armed Conflict and Displacement*, edited by Jo Boyden and Joanna de Berry, 45–62. New York: Bergahn Books.

de la Soudiere, Marie. 2002. "Assessment of the Psychosocial Adjustment of Former Child Soldiers in Sierra Leone: Report to the Andrew W. Mellon Foundation." New York: International Rescue Committee.

DeBurca, Roisin. 2000. "Case Study on Children from the Fighting Forces in Sierra Leone." Document prepared for the International Conference on War-Affected Children,Winnipeg, September. Freetown: UNICEF.

Denov, Myriam. 2010. *Child Soldiers: Sierra Leone's Revolutionary United Front*. Cambridge: Cambridge University Press.

Denzer, LaRay. 1971. "Sierra Leone—Bai Bureh." In *West African Resistance: The Military Response to Colonial Occupation*, edited by Michael Crowder, 233–267. London: Hutchinson & Co.

Denzin, Norman K. 2009. "The Elephant in the Living Room: Or Extending the Conversation about the Politics of Evidence." *Qualitative Research* 9 (2): 139–160.

Diagnostic and Statistical Manual of Mental Disorders (DSM-IV). 1994. 4th ed. Washington, DC: American Psychiatric Association.

Dill, Brian. 2009. "The Paradoxes of Community-Based Participation in Dar es Salaam." *Development and Change* 40 (4): 717–743.

Diouf, Mamadou. 2003. "Engaging Postcolonial Cultures: African Youth and Public Space." *African Studies Review* 46 (1) :1–12.

Dorjahn, V. R. 1960. "The Changing Political System of the Temne." *Africa* 30 (2): 110–39.

Dowdney, Linda. 2007. *Trauma, Resilience and Cultural Healing: How Do We Move Forward?* London: Coalition to Stop the Use of Child Soldiers.

Duffield, Mark 2000. "Globalization, Transborder Trade, and War Economies." In *Greed and Grievance: Economic Agendas in Civil Wars*, edited by Mats Berdal and David M. Malone, 69–90. Boulder, CO: Lynne Rienner.

Durham, Deborah. 2000. "Youth and the Social Imagination in Africa: Introduction to Parts 1 and 2." *Anthropological Quarterly* 73 (3): 113–120.

Ehrenreich, Rosa. 1998. "The Stories We Must Tell: Ugandan Children and the Atrocities of the Lord's Resistance Army." *Africa Today* 45 (1): 79–102.

El-Bushra, Judy. 2003. "Fused in Combat: Gender Relations and Armed Conflict." *Development in Practice* 13 (2-3): 252–265.

Ellis, Stephen. 1999. *The Mask of Anarchy: The Destruction of Liberia and the Religious Dimension of an African Civil War.* New York: New York University Press.

———. 2003. "Young Soldiers and the Significance of Initiation: Some Notes from Liberia." Paper presented at Youth and the Politics of Generational Conflict in Africa, Leiden, the Netherlands, April 24 and 25.

Errante, Antoinette. 2000. "Child Soldiers as Casualties of the Post-Conflict Policy Environment: When is a Country Post-Post-War?" Paper presented at Children and Armed Conflict: Reintegration of Former Child Soldiers in the Post-Conflict Community, Tokyo.

Fanthorpe, Richard. 2001. "Neither Citizen nor Subject? 'Lumpen' Agency and the Legacy of Native Administration in Sierra Leone." *African Affairs* 100:363–386.

Fanthorpe, Richard, and Roy Maconachie. 2010. "Beyond the 'Crisis of Youth'? Mining, Farming, and Civil Society in Post-war Sierra Leone." *African Affairs* 109 (435): 251–272.

Ferme, Mariane. 1998. "The Violence of Numbers: Consensus, Competition, and the Negotiation of Disputes in Sierra Leone." *Cahiers d'Etudes Africaines* 38 (150–152): 555–580.

———. 1999. "Staging *Politisi*: The Dialogics of Publicity and Secrecy in Sierra Leone." In *Civil Society and the Political Imagination in Africa: Critical Perspectives*, edited by John L. Comaroff and Jean Comaroff, 160–191. Chicago: University of Chicago Press.

———. 2001a. "La Figure du chasseur et les chasseurs militiens dans le conflit sierra leonien." *Politique Africaine* 82:119–132.

———. 2001b. *The Underneath of Things: Violence, History, and the Everyday in Sierra Leone.* Berkeley: University of California Press.

Ferme, Mariane, and Daniel Hoffman. 2004. "Hunter Militias and the International Human Rights Discourse in Sierra Leone and Beyond." *Africa Today* 50 (4): 73–95.

Fofanah, Mohamed Pa-Momo. 2004. *Juvenile Justice and Children in Armed Conflict: Facing the Fact and Forging the Future via the Sierra Leone Test*. LLM, Law, Harvard, Cambridge, MA.

Forester, John F. 1999. *The Deliberative Practitioner*. Cambridge, MA: MIT Press.

Fortes, Meyer. 1973. "On the Concept of the Person among the Tallensi." In *La notion de personne en Afrique*, edited by G. Dieterlen, 238–319. Paris: Centre National de la Recherche Scientifique.

Foucault, Michel. 1984a. *L'usage du plaisir. L'histoire de la sexualité*. Vol. 2. Paris: Gallimard.

———. 1984b. *Le souci de soi. L'histoire de la sexualité*, Vol. 3. Paris: Gallimard.

Garfinkel, Harold. 1984[1967]. *Studies in Ethnomethodology*. Cambridge, UK: Polity Press.

Gberie, Lansana. 2005. *A Dirty War in West Africa: The RUF and the Destruction of Sierra Leone*. Bloomington: Indiana University Press.

Gibbs, Sara. 1994. "Post-War Social Reconstruction in Mozambique: Re-framing Children's Experience of Trauma and Healing." *Disasters* 18 (3): 268–276.

Giddens, Anthony. 1979. *Central Problems in Social Theory. Action, Structure and Contradiction in Social Analysis*. London: Macmillan.

———. 1984. *The Constitution of Society*. Cambridge, UK: Polity Press.

Gilligan, Chris. 2009. "'Highly Vulnerable'? Political Violence and the Social Construction of Traumatized Children." *Journal of Peace Research* 46 (1): 119–134.

Goody, Esther N. 1982. *Parenthood and Social Reproduction: Fostering and Occupational Roles in West Africa*. Cambridge Studies in Social Anthropology, vol. 35. Cambridge: Cambridge University Press.

Gottlieb, Alma. 1998. "Do Infants Have Religion? The Spiritual Lives of Beng Babies." *American Anthropologist* 100 (1): 122–135.

Government of Sierra Leone. 1992. *1985 Population and Housing Census of Sierra Leone*. Freetown: Central Statistics Office.

———. 2000. *The Status of Women and Children in Sierra Leone: A Household Survey Report (MICS-2)*. Freetown: Central Statistics Office, Ministry of Development and Economic Planning.

———. 2003. *Sierra Leone National Youth Policy*. Available from http://www.statehouse-sl.org/policies/youth.html.

Green, Edward C., and Alcinda Honwana. 1999. "Indigenous Healing of War Affected Children in Africa" *IK Notes*. Washington, DC: World Bank 10: 1–5.

Grier, Beverly. 2004. "Child Labor and Africanist Scholarship: A Critical Overview." *African Studies Review* 47 (2): 1–25.

Gupta, Akhil. 2001. "Governing Population: The Integrated Child Development Services Program in India." In *States of Imagination: Ethnographic Explorations of the Postcolonial State*, edited by Thomas Blom Hansen and Finn Stepputat, 65–96. Durham, NC: Duke University Press.

Harris, David Alan. 2010. "A 2010 Postscript To: Pathways to Embodied Empathy and Reconciliation after Atrocity: Former Boy Soldiers in A Dance/Movement Therapy

Group in Sierra Leone." Child Soldiers International, Psychosocial Reports. Available at http://www.child-soldiers.org/psychosocial_report_reader.php?id=305.

Henry, Doug. 2000. "Embodied Violence: War and Relief Along the Sierra Leone Border." Ph.D. diss., Southern Methodist University.

Hoffman, Daniel. 2003. "Like Beasts in the Bush: Synonyms of Childhood and Youth in Sierra Leone." *Postcolonial Studies: Culture, Politics, Economy* 6 (3): 295–308.

———. 2005. "The Brookfields Hotel (Freetown, Sierra Leone)." *Public Culture* 17 (1): 55–74.

———. 2011. *The War Machines: Young Men and Violence in Sierra Leone and Liberia.* Durham, NC: Duke University Press.

Honwana, Alcinda. 1997. "Healing for Peace: Traditional Healers and Post-War Reconstruction in Southern Mozambique." *Peace and Conflict: Journal of Peace Psychology* 3 (3): 293–305.

———. 1999. "Negotiating Post-war Identities: Child Soldiers in Mozambique and Angola." *CODESRIA Bulletin,* nos 1 & 2:4–13.

———. 2001. "Children of War: Understanding War and War Cleansing in Mozambique and Angola." In *Civilians in War,* edited by Simon Chesterman, 123–142. Boulder, CO: Lynne Rienner.

———. 2006. *Child Soldiers in Africa.* Philadelphia: University of Pennsylvania Press.

Honwana, Alcinda, and Filip De Boeck. 2005. *Makers & Breakers: Children & Youth in Postcolonial Africa.* Trenton, NJ, and Asmara, Eritrea: Africa World Press.

Howard, Allen M., and David E. Skinner. 1984. "Network Building and Political Power in Northwestern Sierra Leone, 1800–65." *Africa* 54 (2): 2–28.

Human Rights Watch website, http://www.hrw.org/campaigns/crp/index.htm, accessed August 1, 2004

Humphreys, Macartan, and Jeremy M. Weinstein. 2004a. "Handling and Manhandling Civilians in Civil War: Determinants of the Strategies of Warring Factions." Paper presented at Techniques of Violence in Civil War, workshop sponsored by the Centre for Studies of Civil War at the International Peace Research Institute, Oslo, August 20–21.

———. 2004b. "What the Fighters Say: A Survey of Ex-Combatants in Sierra Leone, June–August 2003." Interim report: July 2004." Available online at http://www.columbia.edu/~mh2245/SL.htm.

Ibrahim, Aisha Fofana, and Susan Shepler. 2011. "Introduction to Special Issue: Everyday Life in Postwar Sierra Leone." *Africa Today* 58 (2): v–xii.

Inter-Agency Standing Committee (IASC). 2007. *IASC Guidelines on Mental Health and Psychosocial Support in Emergency Settings.* Geneva: IASC.

Isaac, Barry L., and Shelby R. Conrad. 1982. "Child Fosterage Among the Mende of Upper Bambara Chiefdom, Sierra Leone: Rural-Urban and Occupational Comparisons." *Ethnology* 21 (3): 243–257.

Isiugo-Abanihe, Uche. 1985. "Child Fosterage in West Africa." *Population and Development Review* 11 (1): 53–74.

Jackson, Michael. 1989. *Paths Toward a Clearing: Radical Empiricism and Ethnographic Inquiry.* Bloomington and Indianapolis: Indiana University Press.

———. 2004. *In Sierra Leone*. Durham, NC: Duke University Press.

James, Alison, and Alan Prout. 1997. *Constructing and Reconstructing Childhood: Contemporary Issues in the Sociological Study of Childhood*. London and Washington, DC: Falmer Press.

Jensen, Peter S. and Jon Shaw. 1993. "Children as Victims of War: Current Knowledge and Future Research Needs." *Journal of the American Academy of Child and Adolescent Psychiatry* 32 (4): 697–708.

Jones, Lynne. 2008. "Responding to the Needs of Children in Crisis." *International Review of Psychiatry* 20 (3): 291–303.

Kalyvas, Stathis N. 2006. *The Logic of Violence in Civil War*. Cambridge: Cambridge University Press.

Kandeh, Jimmy. 2002. "Subaltern Terror in Sierra Leone." In *Africa in Crisis: New Challenges and Possibilities*, edited by Tunde Zack-Williams, Diane Frost, and Alex Thomson, 179–195. London: Pluto Press.

Kaplan, Robert. 1994. "The Coming Anarchy: How Scarcity, Crime, Overpopulation, and Disease Are Rapidly Destroying the Social Fabric of Our Planet." *Atlantic Monthly*, February, 44–76.

Keairns, Yvonne. 2002. *The Voices of Girl Child Soldiers*. New York and Geneva: Quaker United Nations Office. Available online at http://www.quno.org/resource/2003/1/voices-girl-child-soldiers.

Keen, David. 2005. *Conflict and Collusion in Sierra Leone*. Oxford: James Currey.

Kelsall, Michelle Staggs, and Shanee Stepakoff. 2007. "'When We Wanted to Talk About Rape': Silencing Sexual Violence at the Special Court for Sierra Leone." *International Journal of Transitional Justice* 1 (3): 355–374.

Kelsall, Tim. 2009. *Culture under Cross-Examination: International Justice and the Special Court for Sierra Leone, Cambridge Studies in Law and Society*. Cambridge, U.K.: Cambridge University Press.

King, Nathaniel. 2012. "Contested Spaces in Post-War Society: The 'Devil Business' in Freetown, Sierra Leone." Ph.D. diss., Martin-Luther-Universität, Halle-Wittenberg, Germany.

Krech, Robert. 2003. "The Reintegration of Former Child Combatants: A Case Study of NGO Programming in Sierra Leone." Master's thesis, University of Toronto.

Lancy, David F. 1996. *Playing on the Mother-Ground: Cultural Routines for Children's Development*. New York, London: Guilford Press.

Latour, Bruno. 1991. *Nous n'avons jamais été modernes. Essai d'anthropologie symétrique*. Paris: La Découverte.

Lave, Jean. 2011. *Apprenticeship in Critical Ethnographic Practice*. Chicago: University of Chicago Press.

Lave, Jean, and Ettiene Wenger. 1991. *Situated Learning: Legitimate Peripheral Participation*. Cambridge: Cambridge University Press.

Le Billon, P. 2003. "Blood Diamonds: Linking Spaces of Exploitation and Regulation." Paper presented at Environmental Politics Workshop, Berkeley, California, October 17.

Leach, Melissa. 2000. "New Shapes to Shift: War, Parks, and the Hunting Person in Modern West Africa." *Journal of the Royal Anthropological Institute* 6: 577–595.

———. 2004. "Introduction to Special Issue: Security, Socioecology, Polity: Mande Hunters, Civil Society, and Nation-States in Contemporary West Africa." *Africa Today* 50 (4): vii–xvi.

Legrand, Jean-Claude. 1997. *Capetown Principles and Best Practices on the Prevention of Recruitment of Children in the Armed Forces and Demobilization and Social Reintegration of Child Soldiers in Africa.* New York: UNICEF.

———. 1999. "Lessons Learned from UNICEF Field Programmes for the Prevention of Recruitment, Demobilization and Reintegration of Child Soldiers." New York: UNICEF.

Levinson, Bradley, Douglas Foley, and Dorothy Holland. 1996. *The Cultural Production of the Educated Person: Critical Ethnographies of Schooling and Local Practice.* Albany: State University of New York Press.

Li, Tania. 2007. *The Will to Improve: Governmentality, Development, and the Practice of Politics.* Durham, NC: Duke University Press.

Liddell, Christine, Jennifer Kemp, and Molly Moema. 1993. "The Young Lions: African Children and Youth in Political Struggle." In *The Psychological Effects of War and Violence on Children*, edited by Lewis A. Leavitt and Nathan A. Fox. Hillsdale, NJ: Lawrence Erlbaum Associates.

Little, Kenneth. 1967. *The Mende of Sierra Leone: A West African People in Transition.* Rev. ed. London: Routledge and K. Paul.

MacCormack, Carol P. 1979. "Sande: The Public Face of a Secret Society." In *The New Religions of Africa*, edited by B. Jules-Rosette, 27–37. Norwood, NJ: Ablex Publishing.

Machel, Graça. 1996. *Impact of Armed Conflict on Children.* New York: United Nations.

MacKenzie, Megan. 2009a. "Empowerment Boom or Bust? Assessing Women's Post-Conflict Empowerment Initiatives." *Cambridge Review of International Affairs* 22 (2): 199–215.

———. 2009b. "Securitization and Desecuritization: Female Soldiers and the Reconstruction of Women in Post-Conflict Sierra Leone." *Security Studies* 18 (2): 241–261.

Malkki, Liisa, and Emily Martin. 2003. "Children and the Gendered Politics of Globalization: In Remembrance of Sharon Stephens." *American Ethnologist* 30 (2): 216–224.

Mann, Gillian. 2004. "Separated Children: Care and Support in Context." In *Children and Youth on the Front Line: Ethnography, Armed Conflict and Displacement*, edited by Jo Boyden and Jo de Berry, 3–22. New York and Oxford: Berghahn Books.

Mansaray, Binta. 2000. "Women Against Weapons: A Leading Role for Women in Disarmament." In *Bound to Cooperate: Conflict Peace and People in Sierra Leone*, edited by Anatole Ayissi and Robin-Edward Poulton, 144–149. Geneva: United Nations Institute for Disarmament Research (UNIDIR).

Marshall, Dominique. 2002. "Humanitarian Sympathy for Children in Times of War and the History of Children's Rights, 1919–1959." In *Children and War: A Historical Anthology*, edited by James Marten, 184–199. New York: New York University Press.

Marten, James. 2002. *Children and War: A Historical Anthology*. New York: New York University Press.

Mattar, Mohamed Y. 2003. "Trafficking in Persons, Especially Women and Children, in Countries of the Middle East: The Scope of the Problem and the Appropriate Legislative Responses." *Fordham International Law Journal* 26:721–760.

Mazurana, Dyan E., Susan A. McKay, Khristopher C. Carlson, and Janel C. Kasper. 2002. "Girls in Fighting Forces and Groups: Their Recruitment, Participation, Demobilization, and Reintegration." *Peace and Conflict: Journal of Peace Psychology* 8 (2): 97–123.

Mazurana, Dyan, and Susan McKay. 2001. "Child Soldiers: What about the Girls?" *Bulletin of the Atomic Scientists* 57 (5): 30–35.

McDermott, Ray. P. 1996. "The Acquisition of a Child by a Learning Disability." In *Understanding Practice: Perspectives on Activity and Context*, edited by Seth Chaiklin and Jean Lave, 269–305. Cambridge: Cambridge University Press.

McEvoy-Levy, Siobhan. 2006. *Troublemakers or Peacemakers? Youth and Post-Accord Peace Building*, Notre Dame, IN: University of Notre Dame Press.

McKay, Susan. 1998. "The Effects of Armed Conflict on Girls and Women." *Peace and Conflict: Journal of Peace Psychology* 4 (4): 381–392.

———. 2000. "Gender Justice and Reconciliation." *Women's Studies International Forum* 23 (5): 561–570.

———. 2004. "Reconstructing Fragile Lives: Girls' Social Reintegration in Northern Uganda and Sierra Leone." *Gender and Development* 12 (3): 19–30.

———. 2006. "Girlhoods Stolen: The Plight Of Girl Soldiers During and After Armed Conflict." In *A World Turned Upside Down: Social Ecological Approaches to Children in War Zones*, edited by Neil Boothby, Alison Strang and Michael Wessells, 89–110. Bloomfield, CT: Kumarian Press.

McKay, Susan, Mary Burman, Maria Gonsalves, and Miranda Worthen. 2004. "Known but Invisible: Girl Mothers Returning from Fighting Forces." *Coalition to Stop the Use of Child Soldiers Newsletter* Issue 11.

McKay, Susan, and Dyan Mazurana. 2004. *Where are the Girls? Girls in Fighting Forces in Northern Uganda, Sierra Leone and Mozambique: Their Lives During and After War*. Montreal: Rights and Democracy.

McKay, Susan, Angela Veale, Miranda Worthen, and Mike Wessells. 2010. "Community-Based Reintegration of War-Affected Young Mothers: Participatory Action Research (PAR) in Liberia, Sierra Leone & Northern Uganda." University of Wyoming. All project materials available online at http://www.uwyo.edu/girlmotherspar/.

McNaughton, Patrick R. 1988. *The Mande Blacksmiths: Knowledge, Power, and Art in West Africa*. Bloomington: Indiana University Press.

Medeiros, Emilie. 2007. "Integrating mental health into Post-Conflict Rehabilitation. The Case of Sierra Leonean and Liberian 'Child Soldiers.'" *Journal of Health Psychology* 12 (3): 498–504.

Merry, Sally Engle. 2006. "Transnational Human Rights and Local Activism: Mapping the Middle." *American Anthropologist* 108 (1): 38–51.

Miettinen, Reijo, Dalvir Samra-Fredericks, and Dvora Yanow. 2009. "Re-Turn to Practice: An Introductory Essay." *Organization Studies* 30 (12): 1309–1327.

Moran, Mary. 1994. "Warriors or Soldiers: Masculinity and Ritual Transvestism in the Liberian Civil War." In *Feminism, Nationalism, and Militarism*, edited by Constance Sutton, 73–88. Arlington, VA.: American Anthropological Association/Association for Feminist Anthropology.

Muana, Patrick. 1997. "The Kamajoi Militia: Violence, Internal Displacement and the Politics of Counter Insurgency." *Africa Development* 22 (3/4): 77–100.

Murphy, William. 1980. "Secret Knowledge as Property and Power in Kpelle Society: Elders versus Youth." *Africa* 50:193–207.

———. 2003. "Military Patrimonialism and Child Soldier Clientelism in the Liberian and Sierra Leonean Civil Wars." *African Studies Review* 46 (2): 61–87.

———. 2010. "Right to Exit a Community (And Not Return): Youth and Women Critique of Clientelist Dependency in Post-Conflict Liberia and Sierra Leone." Conference paper at the Upper Guinea Coast in Transnational Perspective, Max Planck Institute for Social Anthropology, Halle, Germany. December 9.

National Forum for Human Rights. 2001. *Annual Report 2001*. Freetown: National Forum for Human Rights.

Nieuwenhuys, Olga. 1996. "The Paradox of Child Labor and Anthropology." *Annual Review of Anthropology* 25: 237–251.

Nordstrom, Carolyn. 1997. *A Different Kind of War Story*. Philadelphia: University of Pennsylvania Press.

———. 2004. *Shadows of War: Violence, Power, and International Profiteering in the Twenty-First Century*. Berkeley: University of California Press.

Nunley, John W. 1987. *Moving with the Face of the Devil: Art and Politics in Urban West Africa*. Urbana: University of Illinois Press.

O'Brien, Donal Cruise. 1996. "A Lost Generation? Youth Identity and State Decay in West Africa." In *Postcolonial Identities in Africa*, edited by Richard Werbner and Terence Ranger, 55–74. London: Zed Books.

Park, Augustine S.J. 2006. "'Other Inhumane Acts': Forced Marriage, Girl Soldiers and the Special Court for Sierra Leone." *Social & Legal Studies* no. 15 (3):315–337.

Peace Agreement Between the Government of Sierra Leone and the Revolutionary United Front of Sierra Leone, Lomé, Togo. 1999. Available from http://www.usip.org/library/pa/sl/sierra_leone_07071999_toc.html.

Pederson, Jon. 2001. "What Should We Know about Children in Armed Conflict and How Should We Go about Knowing It?" Paper presented at Filling Knowledge Gaps: A Research Agenda on the Impact of Armed Conflict on Children, Florence, July 2–4.

Peters, Krijn. 2000. "Policy Making on Children in Conflict: Lessons from Sierra Leone and Liberia." *Cultural Survival Quarterly* 24 (2): 56.

———. 2006. "Footpaths to Reintegration: Armed Conflict, Youth and the Rural Crisis in Sierra Leone." Ph.D. diss., Wageningen University, Wageningen, the Netherlands.

———. 2007. "Reintegration Support for Young Ex-Combatants: A Right or a Privilege?" *International Migration* 45 (5): 35–59.

———. 2011. *War and the Crisis of Youth in Sierra Leone.* London: International African Institute and Cambridge University Press.

Peters, Krijn and Paul Richards. 1998. "'Why We Fight': Voices Of Youth Combatants in Sierra Leone." *Africa,* 68 (2): 183–210.

Pupavac, Vanessa. 2001. "Misanthropy without Borders: The International Children's Rights Regime." *Disasters* 25 (2): 95–112.

Reckwitz, Andreas. 2002. "Toward a Theory of Social Practices: A Development in Culturalist Theorizing." *European Journal of Social Theory* 5 (2): 243–263.

Reno, William. 1997a. "Privatizing War in Sierra Leone." *Current History* 96 (610): 227.

———. 1997b. "War, Markets, and the Reconfiguration of West Africa's Weak States." *Comparative Politics* 29 (4): 493–510.

———. 1998. *Warlord Politics and African States.* Boulder: Lynne Rienner.

Richards, Paul. 1995. "Rebellion in Liberia and Sierra Leone: A Crisis of Youth?" In *Conflict in Africa,* edited by Oliver Furley, 134–170. London: Tauris.

———. 1996. *Fighting for the Rain Forest: War, Youth, and Resources in Sierra Leone.* Oxford: James Currey.

———. 2009. "Dressed to Kill: Clothing as Technology of the Body in the Civil War in Sierra Leone." *Journal of Material Culture* 14 (4): 495–512.

Richards, Paul, Steven Archibald, Khadija Bah, and James Vincent. 2003. "Where Have All the Young People Gone? Transitioning Ex-Combatants Towards Community Reconstruction after the War in Sierra Leone." Unpublished report submitted to the National Commission for Disarmament, Demobilisation and Reintegration, Government of Sierra Leone.

Richards, Paul, Khadija Bah, and James Vincent. 2004. "Social Capital and Survival: Prospects for Community-Driven Development in Post-Conflict Sierra Leone." Social Development Papers: Community-Driven Development,/Conflict Prevention and Reconstruction, paper no. 12. Washington, DC: World Bank.

Riley, Stephen. 1996. *Liberia and Sierra Leone: Anarchy or Peace in West Africa?* London: Research Institute for the Study of Conflict and Terrorism.

———. 1997. "Sierra Leone: The Militariat Strikes Again." *Review of African Political Economy* 72: 287–292.

Riley, Steve, and Max Sesay. 1995. "Sierra Leone: The Coming Anarchy?" *Review of African Political Economy* 63:121–126.

Rosen, David. 2005. *Armies of the Young: Child Soldiers in War and Terrorism.* New Brunswick, NJ: Rutgers University Press.

Save the Children UK. 2002. "Extensive Abuse of West African Refugee Children Reported." Press Release. London: Save the Children UK.

Schafer, Jessica. 2004. "The Use of Patriarchal Imagery in the Civil War in Mozambique and Its Implications for the Reintegration of Child Soldiers." In *Children and Youth on the Front Line: Ethnography, Armed Conflict and Displacement*, edited by Jo Boyden and Joanna de Berry, 87–104. New York: Berghahn Books.

Schildkrout, Enid. 1973. "The Fostering of Children in Urban Ghana: Problems of Ethnographic Analysis." *Urban Anthropology* 2 (1): 48–73.

Schroven, Anita. 2005. "Choosing between Different Realities: Gender Mainstreaming and Self-Images of Women After Armed Conflict in Sierra Leone." M.A. thesis, Georg-August-Universität, Gottingen.

Shaw, Rosalind. 2002. *Memories of the Slave Trade: Ritual and the Historical Imagination in Sierra Leone*. Chicago: University of Chicago Press.

———. 2003. "Robert Kaplan and 'Juju Journalism' in Sierra Leone's Rebel War: The Primitivizing of an African Conflict." In *Magic and Modernity: Interfaces of Revelation and Concealment*, edited by Birgit Meyer and Peter Pels, 81–102. Stanford, CA: Stanford University Press.

———. 2005. "Rethinking Truth and Reconciliation Commissions: Lessons from Sierra Leone." Washington, DC: United States Institute of Peace.

Shepler, Susan. 1998. "Education as a Site of Political Struggle in Sierra Leone." *Antropológicas* 2: 3–14.

———. 2002. "Les Filles-Soldats: Trajectoires d'apres-guerre en Sierra Leone." *Politique Africaine* 88: 49–62.

———. 2003. "Educated in War: The Rehabilitation of Child Soldiers in Sierra Leone." In *Conflict Resolution and Peace Education in Africa*, edited by Ernest Uwazie, 57–76. Lanham, MD: Lexington Books.

———. 2004. "Globalizing Child Soldiers in Sierra Leone." In *Youthscapes: The Popular, The National, The Global*, edited by Sunaina Maira and Elisabeth Soep, 119–133. Philadelphia: University of Pennsylvania Press.

———. 2005. "The Rites of the Child: Global Discourses of Youth and Reintegrating Child Soldiers in Sierra Leone." *Journal of Human Rights* 4 (2): 197–211.

———. 2010a. "Are "Child Soldiers" in Sierra Leone a New Phenomenon?" In *The Powerful Presence of the Past: Integration and Conflict along the Upper Guinea Coast*, edited by Jacqueline Knörr and Wilson Trajano Filho, 297–321. Leiden, the Netherlands: Brill Publishers.

———. 2010b. "Shifting Priorities in Child Protection in Sierra Leone since Lomé." In *Sierra Leone Beyond Lomé: Challenges and Possibilities for a Post-War Nation*, edited by Marda Mustapha and Joseph Bangura, 69–82. New York: Palgrave Macmillan.

———. 2010c. "Youth Music and Politics in Post-War Sierra Leone." *Journal of Modern African Studies* 48 (4): 627–642.

Siddle, D. J. 1968. "War-Towns in Sierra Leone: A Study in Social Change." *Africa* 38 (1): 47–56.

Sierra Leone Truth and Reconciliation Commission. 2004. "Sierra Leone Truth and Reconciliation Report." Freetown.

Singer, Peter W. 2005. *Children at War*. New York: Pantheon Books.

Smillie, Ian, Lansana Gberie, and Ralph Hazleton. 2000. *The Heart of the Matter: Sierra Leone, Diamonds & Human Security*. Ottawa: Partnership Africa Canada.

Sommers, Marc. 2007. "Embracing the Margins: Working with Youth amid War and Insecurity." In *Too Poor for Peace? Global Poverty, Conflict, and Security in the 21st Century*, edited by Lael Brainard and Derek Chollet, 101–118. Washington, DC: Brookings Institution Press.

Stark, Lindsay, Alastair Ager, Mike Wessells, and Neil Boothby. 2009a. "Developing Culturally Relevant Indicators of Reintegration for Girls, Formerly Associated with Armed Groups, in Sierra Leone Using a Participative Ranking Methodology." *Intervention* 7 (1): 4–16.

Stark, Lindsay, Neil Boothby, and Alastair Ager. 2009b. "Children and Fighting Forces: 10 Years on from Cape Town." *Disasters* 33 (4): 522–547.

Stavrou, Vivi. 2004. *Breaking the Silence: Girls Forcibly Involved in the Armed Struggle in Angola*. Luanda: Christian Children's Fund Angola, CIDA Child Protection Research Fund.

Stephens, Sharon. 1995. *Children and the Politics of Culture*. Princeton, NJ: Princeton University Press.

Stoler, Ann Laura. 1995. *Race and the Education of Desire: Foucault's History of Sexuality and the Colonial Order of Things*. Durham, NC: Duke University Press.

Tangri, Roger. 1976. "Conflict and Violence in Contemporary Sierra Leone Chiefdoms." *Journal of Modern African Studies* 14 (2): 311–321.

Thompson, Carol B. 1999. "Beyond Civil Society: Child Soldiers as Citizens in Mozambique." *Review of African Political Economy* 26 (80): 191–206.

United Nations Children's Fund [UNICEF]. 2007. *The Paris Principles. Principles and Guidelines on Children Associated With Armed Forces or Armed Groups*. February.

United Nations Development Programme [UNDP]. 2003. *Human Development Report 2003. Millennium Development Goals: A Compact among Nations to end Human Poverty*. New York: UNDP.

———. 2006. *Youth and Violent Conflict: Society and Development in Crisis?* New York: UNDP.

Urdal, Henrik. 2004. "The Devil in the Demographics: The Effect of Youth Bulges on Domestic Armed Conflict, 1950–2000." Social Development Papers/Conflict Prevention and Reconstruction, paper no. 14. Washington, DC: World Bank.

Utas, Mats. 2003. "Sweet Battlefields: Youth and the Liberian Civil War." Ph.D. diss., Uppsala University.

———. 2004. "Fluid Research Fields: Studying Excombatant Youth in the Aftermath of the Liberian Civil War." In *Children and Youth on the Front Line: Ethnography, Armed Conflict and Displacement*, edited by Jo Boyden and Joanna de Berry, 209–236. New York: Berghahn Books.

———. 2005a. "Agency of Victims: Young Women in the Liberian Civil War." In *Makers & Breakers: Children & Youth in Postcolonial Africa*, edited by Alcinda

Honwana and Filip de Boeck, 53–80. Trenton, NJ, and Asmara, Eritrea: Africa World Press.

———. 2005b. "Building a Future? The Reintegration and Remarginalisation of Youth in Liberia." In *No Peace, No War: And Anthropology of Contemporary Armed Conflicts*, edited by Paul Richards, 137–154. Athens: Ohio University Press.

———. 2005c. "Victimcy, Girlfriending, Soldiering: Tactic Agency in a Young Woman's Social Navigation of the Liberian War Zone." *Anthropological Quarterly* 78 (2): 403–430.

Verhoef, Heidi. 2005. "'A Child Has Many Mothers': Views of Child Fostering in Northwestern Cameroon." *Childhood* 12 (3): 369–390.

Vigh, Henrik. 2006. *Navigating Terrains of War: Youth and Soldiering in Guinea-Bissau*. New York and Oxford: Berghahn Books.

Wang, Lianqin. 2007. *Education in Sierra Leone: Present Challenges, Future Opportunities, Africa Human Development Series*. Washington, DC: World Bank.

Wessells, Michael. 2006. *Child Soldiers: From Violence to Protection*. Cambridge: Harvard University Press.

———. 2007. *The Recruitment and Use of Girls in Armed Forces and Groups in Angola: Implications for Ethical Research*. Pittsburgh: Ford Institute for Human Security.

———. 2009. "Supporting the Mental Health and Psychosocial Well-Being of Former Child Soldiers." *Journal of the American Academy of Child and Adolescent Psychiatry* 48(6): 587–590.

Wessells, Michael, and Carlinda Monteiro. 2000. "Healing Wounds of War in Angola: A community-based approach." In *Addressing Childhood Adversity*, edited by David Donald, Andrew Dawes and Johann Louw, 176–201. Cape Town: David Philip Publishers.

West, Harry G. 2000. "Girls with Guns: Narrating the Experience of War of Frelimo's 'Female Detachment.'" *Anthropological Quarterly* 73 (4): 180–194.

Williamson, John. 2005. "Reintegration of Child Soldiers in Sierra Leone." USAID, Displaced Children and Orphan's Fund.

———. 2006. "The Disarmament, Demobilization and Reintegration of Child Soldiers: Social and Psychological Transformation in Sierra Leone." *Intervention* 4 (3): 185–205.

Wlodarczyk, Nathalie. 2009. *Magic and Warfare: Appearance and Reality in Contemporary African Conflict and Beyond*. New York: Palgrave Macmillan.

Women's Commission for Refugee Women and Children. 2000. *Untapped Potential: Adolescents affected by armed conflict, A review of programs and policies*. New York: Women's Commission for Refugee Women and Children.

———. 2004. *Global Survey on Education in Emergencies*. New York: Women's Commission for Refugee Women and Children.

Young, Allan. 1995. *The Harmony of Illusions: Inventing Post-Traumatic Stress Disorder*. Princeton, NJ: Princeton University Press.

Zack-Williams, Alfred. 1999. "Sierra Leone: the Political Economy of Civil War, 1991–98." *Third World Quarterly* 20 (1): 143–162.

———. 2001. "Child Soldiers in the Civil War in Sierra Leone." *Review of African Political Economy* 87:73–82.

Small arms, 22–23
"Sobels," 11, 103
Social practice, 60, 83, 90, 96, 130, 156, 157, 158
Social practice theory, 5–7, 167n7
South Africa, 61, 172n21, 181n10
Special Court for Sierra Leone (SCSL), 13, 61, 63
Spontaneous reintegration, 17, 84, 103, 105. *See also* Informal reintegration
Stephens, Sharon, 7
Stigma, 124, 129, 150, 152, 163
Stoler, Ann, 180n1
Strategy, 5, 7, 38, 72, 83–84, 90, 104, 130, 159; children's strategies regarding ICCs, 85, 87; by communities, 124–128; across contexts, 90; whether to demobilize as a child or an adult, 109–110, 141; secrecy as a strategy, 17, 150, 179n20; strategic self presentation, 83, 99, 145–146, 155–156, 163; strategic use of child rights, 155
Street children, 23, 61, 64, 115, 118, 137
Structural violence, 18, 96; gender inequality as structural violence, 152
Structure and agency, 6–7, 158

Tangri, Roger, 48
Taylor, Charles, 13
Tetherball, 56, 70, 86
Tradition, 13, 26, 53, 72, 132–134, 155; and the CDF, 13, 133, 145
Traditional healing, 71–72, 174n15, 174n16, 175n1
Trajectories; of child soldiers, 79, 83–84; of girls, 150, 151; postwar, 106, 118, 145, 150–151
Trauma, 68, 87–90, 111, 118–119, 155, 163; 173n11; social aspects of, 69–70
Truth and Reconciliation Commissions, 13, 60, 63. *See also* Sierra Leone Truth and Reconciliation Commission

Uganda, 63, 170n2, 174n16, 177n6
Unaccompanied children, 23, 123

United Nations Children's Fund (UNICEF), 3, 15, 22, 60, 61, 62, 73, 97, 109, 127, 167n2, 172n20, 176n9; in the Lomé Peace Agreement, 61
United Nations Convention on the Rights of the Child (UNCRC), 5. *See also* Convention on the Rights of the Child
United Nations Development Programme (UNDP), 60
United Nations High Commissioner for Refugees (UNHCR), 1–2, 16, 60, 153
United Nations Mission in Sierra Leone (UNAMSIL), 13, 108, 110, 117, 121, 153, 173n4, 178n10, 178n12; UNAMSIL child protection officer, 62
United Nations Security Council Resolution 1325 on Women, Peace, and Security, 149
United Nations Special Representative of the Secretary General (SRSG) on Children and Armed Conflict, 61
United States Agency for International Development (USAID), 71, 173n3, 174n18
Utas, Mats, 151, 176n4

Vernacularization of rights, 14, 99, 132, 157
Violent youth, as a historical category, 46

"War boys," 48
War Child (child protection NGO), 62, 69
War economy, 12
War in Sierra Leone, xi–xii, 8–10; fighting factions, 10–14
Wessells, Michael, 70, 71, 174n15
West, Harry, 163
Western model of childhood and youth, 6, 21, 53, 120, 129, 154, 158, 160, 162, 165, 167n3, 169n1
Williamson, John, 71, 173n3
World Bank, 45, 71, 74
World Food Programme (WFP), 60

Youth, 9–10, 25–29; in Africa, 10; definition of, 25, 28–29; as a political category, 28; as a shifter, 25
Youth bulge, 9, 22

ABOUT THE AUTHOR

Susan Shepler is Associate Professor of International Peace and Conflict Resolution in the School of International Service at American University in Washington, DC. She teaches courses on youth and conflict in Africa. Her work has appeared in the *Journal of Modern African Studies, Africa Today, Anthropology Today,* and the *Journal of Human Rights.*

Printed in the United States
by Baker & Taylor Publisher Services